Paramedics! Test Yourself in Anatomy and Physiology

T0175981

Paramedics!
Test Yourself in
Anatomy and Physiology

**Katherine M.A. Rogers,
William N. Scott, Stuart Warner
and Bob Willis**

 Open University Press

Open University Press
McGraw-Hill Education
McGraw-Hill House
Shoppenhangers Road
Maidenhead
Berkshire
England
SL6 2QL

email: enquiries@openup.co.uk
world wide web: www.openup.co.uk

and Two Penn Plaza, New York, NY 10121-2289, USA

First Published 2011

A catalogue record of this book is available from the British Library

ISBN-13: 978-0-33-524370-9
ISBN-10: 0-33-524370-3
eISBN: 978-0-33-524371-6

Library of Congress Cataloging-in-Publication Data
CIP data applied for

Illustrations by Gary Holmes
Typeset in Bell Gothic by RefineCatch Limited, Bungay, Suffolk
Printed and bound by CPI Group (UK) Ltd, Croydon, CR0 4YY

The *McGraw-Hill* Companies

Contents

Using this book

Welcome to *Paramedics! Test Yourself in Anatomy and Physiology*. We hope you will find this an invaluable tool throughout your studies, and beyond!

This book is designed to be used as a revision aid that you can use with your main textbook. Each chapter is designed for stand-alone revision, meaning that you need not read from the beginning to benefit from the book.

Every chapter begins with a brief introduction covering the main points of the topic and directing you to some useful resources. The chapter progresses, providing you with different types of questions that help you test your knowledge of the area. These are:

- *Labelling Exercise*: identify the different elements on the diagram.
- *True or False*: identify if the statement is true or false.
- *Multiple Choice*: identify which of four answers is correct.
- *Fill in the Blanks*: fill in the blanks to complete the statement.
- *Match the Terms*: identify which term matches which statement.

The questions have been designed to be slightly more challenging in each section. Do not ignore a question type just because you are not examined in that way, because the answer will contain useful information that could easily be examined in an alternative question format.

Answers are also provided in each chapter with detailed explanations – this is to help you with revision but can also be used as a learning aid.

We have suggested some useful textbooks that may be used to support your recommended text, but they should not replace the core reading for your course.

A list of directional terms and a list of common prefixes and suffixes used in anatomy and physiology are provided. At the back of the book you will find a glossary.

We hope that you enjoy using this book and that you find it a convenient and useful tool throughout your studies!

GUIDE TO TEXTBOOK RESOURCES

Paramedics! Test Yourself in Pathophysiology
Katherine M.A. Rogers, William N. Scott, Stuart Warner and Bob Willis
Published by McGraw-Hill, 2011

Principles of Anatomy and Physiology (12th edition): Volumes I and II
Gerard J. Tortora and Bryan H. Derrickson
Published by John Wiley & Sons, 2011

Mader's Understanding Human Anatomy and Physiology (6th edition)
Susannah N. Longenbaker
Published by McGraw-Hill, 2010

Hole's Essentials of Human Anatomy and Physiology (11th edition)
David Shier, Jackie Butler and Ricki Lewis
Published by McGraw-Hill, 2011

Ross and Wilson's Anatomy and Physiology in Health and Illness (11th edition)
Anne Waugh and Allison Grant
Published by Churchill Livingstone, Elsevier, 2010

Anatomy and Physiology (8th edition)
Rod R. Seeley, Trent D. Stephens and Philip Tate
Published by McGraw-Hill, 2007

List of abbreviations

These common abbreviations are used in the clinical setting and throughout this book.

ACE	angiotensin-converting enzyme	GnRH	gonadotropin-releasing hormone
ACTH	adrenocorticotrophic hormone	HCG	human chorionic gonadotropin
ADH	antidiuretic hormone (or vasopressin)	HCl	hydrochloric acid
		HR	heart rate
ADP	adenosine diphosphate	Ig	immunoglobulin
ANS	autonomic nervous system	LH	luteinizing hormone
ATP	adenosine triphosphate	MI	myocardial infarction (or heart attack)
AV	atrioventricular		
BP	blood pressure	O_2	oxygen
CF	cystic fibrosis	PEFR	peak expiratory flow rate
CNS	central nervous system	PNS	peripheral nervous system
CO_2	carbon dioxide	PTH	parathyroid hormone
COPD	chronic obstructive pulmonary disease	RAS	renin-angiotensin system
		RBC	red blood cell
CSF	cerebrospinal fluid	RNA	ribonucleic acid
CVA	cerebrovascular accident (or stroke)	SA	sinoatrial
		SER	smooth endoplasmic reticulum
DCT	distal convoluted tubule		
DNA	deoxyribonucleic acid	STI	sexually transmitted infection
EC	enterochromaffin		
ECG	electrocardiogram	SV	stroke volume
ER	endoplasmic reticulum	TSH	thyroid-stimulating hormone
FSH	follicle-stimulating hormone	UV	ultra-violet
GABA	gamma aminobutyric acid	WBC	white blood cell
GI	gastrointestinal		

Directional terms

Abduct   move away from the midline of the body; the opposite of adduct

Adduct   movement towards the midline of the body; the opposite of abduct

Anterior   front-facing or ventral; opposite of posterior or dorsal

Contralateral   on opposite side; opposite of ipsilateral

Distal   far away from point of origin; the opposite of proximal

Dorsal   to the back or posterior of; opposite of ventral or anterior

Inferior   lower or beneath; opposite of superior

Ipsilateral   on same side; opposite of contralateral

Lateral   referring to the side, away from the midline; opposite of medial

Medial   towards the middle; opposite of lateral

Posterior   back or dorsal; opposite of anterior or ventral

Proximal   nearest to the centre of the body; opposite of distal

Superior   above or higher; opposite of inferior

Ventral   referring to front or anterior; opposite of dorsal or posterior

The table below summarizes how these terms match up.

Direction	Opposite term
Abduct	Adduct
Anterior/ventral	Posterior/dorsal
Contralateral	Ipsilateral
Distal	Proximal
Inferior	Superior
Lateral	Medial

Common prefixes, suffixes and roots

Prefix/suffix/ root	Definition	Example
a-/an-	deficiency, lack of	*anuria = decrease or absence of urine production*
-aemia	of the blood	*ischaemia = decreased blood supply*
angio-	vessel	*angiogenesis = growth of new vessels*
broncho-	bronchus	*bronchitis = inflammation of the bronchus*
card-	heart	*cardiology = study of the heart*
chole-	bile or gall bladder	*cholecystitis = inflammation of gall bladder*
cyto-	cell	*cytology = study of cells*
derm-	skin	*dermatology = study of the skin*
entero-	intestine	*enteritis = inflammation of the intestinal tract*
erythro-	red	*erthyropenia = deficiency of red blood cells*
gast-	stomach	*gastritis = Inflammation of stomach lining*
-globin	protein	*haemoglobin = iron-containing protein in the blood*
haem-	blood	*haemocyte = a blood cell (especially red blood cell)*
hepat-	liver	*hepatitis = inflammation of the liver*
-hydr-	water	*rehydrate = replenish body fluids*
leuco-	white	*leucopenia = deficiency of white blood cells*
lymph-	lymph tissue/vessels	*lymphoedema = fluid retention in lymphatic system*
myo-	muscle	*myocardium = cardiac muscle*
nephr-	kidney	*nephritis = inflammation of the kidneys*
neuro-	nerve	*neurology = study of the nerves*
-ology	study of	*histology = study of the tissues*
-ophth-	eye	*ophthalmology = study of the eyes*
osteo-	bone	*osteology = study of bones*
path-	disease	*pathology = study of disease*
pneumo-	air/lungs	*pneumonitis = inflammation of lung tissue*
-uria	urine	*haematuria = blood in the urine*
vaso-	vessel	*vasoconstriction = narrowing of vessels*

1 The human cell

INTRODUCTION

All living things are composed of cells, which are the smallest units of life and are so small they can only be viewed through a microscope. Cells are made from pre-existing cells through cell replication and division. The human body is composed of billions of cells which are specially adapted for their various functions, for example, neural, blood and skin cells.

Despite their functional differences, cells generally possess similar structures and perform similar biochemical processes. Groups of similar cells are arranged into tissues to perform a particular function. Different tissues are organized into organ structures and several organs may be linked together to perform major biological functions in an organ system. Cooperation between the organ systems is essential for maintaining a normal healthy body.

It is important for a paramedic to know that the cells are the building blocks of tissues and organs and how essential molecules are transported in and out of cells, because when the growth or transport mechanisms are disrupted and homeostasis is lost, disease or illness can develop.

> **Useful resources**
>
> Paramedics! Test Yourself in Pathophysiology
>
> Chapters 1 and 2
>
> Mader's Understanding Human Anatomy and Physiology
>
> Chapters 1, 3 and 4
>
> WebAnatomy – Cells, histology:
>
> http://msjensen.cehd.umn.edu/webanatomy/default.htm

LABELLING EXERCISE

1–13 Identify the features of the human cell in Figure 1.1, using the options provided in the box below.

nucleolus	cytoplasm
ribosomes	smooth ER
nuclear membrane	Golgi apparatus
rough ER	lysosome
mitochondrion	nucleus
vacuole	centriole
cell membrane	

Figure 1.1 The human cell

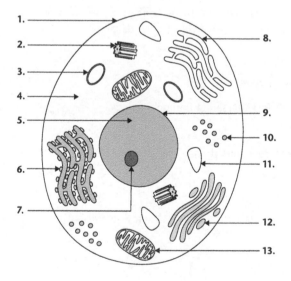

TRUE OR FALSE?

Are the following statements true or false?

14 Cell membranes are fully permeable.

15 Organelles are membrane-bound structures found in the cytoplasm.

16 The cytoskeleton is a lattice structure made of calcium that gives shape to cells.

17 There are three classes of membrane in the human body.

18 There are three major types of tissue in the human body.

19 A gland is a single cell or group of cells adapted for secretion.

20 There are 10 major body systems.

MULTIPLE CHOICE

Identify one correct answer for each of the following.

21 Cellular respiration is defined as:

a) an intracellular, energy-producing process
b) an extracellular, energy-producing process
c) an intracellular, energy-requiring process
d) an extracellular, energy-requiring process

22 Cellular diffusion is where solutes:

a) move from a low concentration to a high concentration
b) move from a high concentration to a low concentration
c) stay in their relative chambers, due to other pressures within the body
d) stay in their relative chambers, unaffected by other pressures within the body

23 Osmosis describes the process when water:

a) moves across a membrane from an area of higher concentration (with lower solutes) to lower concentration (with higher solutes)
b) moves across a membrane from an area of lower concentration (with higher solutes) to higher concentration (with lower solutes)
c) stays in its relative chamber, unaffected by dissolved solutes
d) stays in its relative chamber, due to the presence of dissolved solutes

24 Phagocytosis is where cells:

   a) ingest and destroy microbes, cell debris and other foreign matter
   b) export substances from the cell to the extracellular space
   c) move fluid from the extracellular space to within the cell
   d) produce erythrocytes for release within the systemic system

25 Prokaryotes lack:

   a) a cell membrane
   b) a ribosome
   c) cytoplasm
   d) a cell nucleus

FILL IN THE BLANKS

Fill in the blanks in each statement using the options in the box below.
Not all of them are required, so choose carefully!

histology	nucleus
physiology	lysosome
cytology	pathology
pathophysiology	connective
organelles	mitochondria
anatomy	

26 The science of body structure is called _____; the study of body function is _____.

27 The _____ are the 'powerhouses' of the cell because they produce energy in the form of ATP.

28 The genetic material (DNA) is found in the _____.

29 _____ are structures within cells that perform cellular functions.

30 The cellular organelle responsible for digesting exhausted cell components is the _____.

31 The study of cells is called _____; the study of tissues is called _____.

32 Blood is a type of _____ tissue.

ANSWERS

 LABELLING EXERCISE

Figure 1.2 The human cell

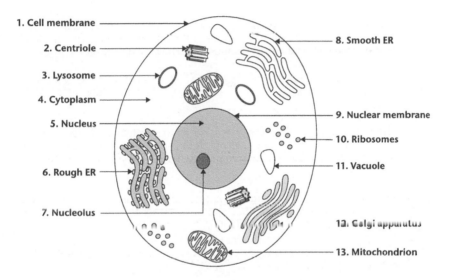

1. Cell membrane
2. Centriole
3. Lysosome
4. Cytoplasm
5. Nucleus
6. Rough ER
7. Nucleolus
8. Smooth ER
9. Nuclear membrane
10. Ribosomes
11. Vacuole
12. Golgi apparatus
13. Mitochondrion

1 *Cell (or plasma) membrane:* surrounds the cell, separating the external environment from the cytoplasm and controlling what enters and leaves the cell. Cell membranes are comprised of a double layer of phospholipid molecules tightly packed together. The membrane is partially permeable and has protein molecules embedded in it that allow transport of substances in and out of the cell.

2 *Centriole:* small structure located close to the nuclear membrane involved in cell division.

3 *Lysosome:* formed from the ends of the Golgi apparatus. Contains digestive enzymes, collectively called lysozyme, which is used by phagocytes to digest bacteria and is responsible for autolysis of cells after their death. Lysozyme is found in tears and helps protect the eye against bacterial infection.

7

4 *Cytoplasm (or protoplasm):* a gel-like substance that surrounds all organelles outside the nucleus. It also contains enzymes that speed up (catalyze) biochemical reactions in the cell.

5 *Nucleus:* contains the genetic material, deoxyribonucleic acid (DNA). DNA is organized into genes on chromosomes that control protein synthesis.

6 *Rough endoplasmic reticulum (RER):* system of tubes and sacs studded with ribosomes on the external surface. Ribosomes are involved in the manufacture of membrane-bound proteins.

7 *Nucleolus:* round structure in the nucleus where ribosomes are produced. Some nuclei have more than one nucleolus. The nucleoli contain large amounts of DNA and ribonucleic acid (RNA).

8 *Smooth endoplasmic reticulum (SER):* system of tubes and sacs in which lipids are synthesized (SER does not have ribosomes on its surface).

9 *Nuclear membrane:* barrier separating the nuclear contents from the cytoplasm. There are numerous pores in the nuclear membrane, which facilitate and regulate the exchange of materials between the nucleus and the cytoplasm.

10 *Ribosomes:* responsible for manufacturing proteins. Some are found unattached in the cytoplasm but more commonly attached to the surface of rough endoplasmic reticulum (hence the name rough ER).

11 *Vacuole:* small organelle with a minor function facilitating transport in and out of cells (endocytosis and exocytosis).

12 *Golgi apparatus:* flattened sacs that package proteins and carbohydrates into vesicles for export from the cell.

13 *Mitochondrion (pl. mitochondria):* sometimes called the cell's 'powerhouse' because it produces energy in the form of adenosine triphosphate (ATP), which provides energy for muscle contraction, active transport and to build large molecules.

TRUE OR FALSE?

14 **Cell membranes are fully permeable.**

Cell membranes are selectively (or partially) permeable allowing only certain substances to enter the cell. The membranes of cell organelles are also selectively permeable.

15 **Organelles are membrane-bound structures found in the cytoplasm.**

Organelles are small membrane-bound structures found in the cytoplasm of the cell that have a specific function. Some organelles are double-membrane structures (such as mitochondria) whereas other organelles are enclosed in a single membrane (for example, lysosomes).

16 **The cytoskeleton is a lattice structure made of calcium that gives shape to cells.**

The cytoskeleton is a complex lattice structure of *protein* microtubules that maintains cell shape.

17 **There are three classes of membrane in the human body.**

There are *four* classes of membrane: *mucous, cutaneous, serous* and *synovial*. Membranes are sheet-like structures found throughout the body and perform specific functions. Mucous membranes line openings to the body such as the respiratory, digestive, urinary and reproductive tracts. The cells of this tissue produce lubricating mucus. Cutaneous membrane lines the skin. Serous membranes produce serous fluid which reduces friction between tissues and organs; it has two layers, the parietal and visceral. Parietal layers line the walls of cavities in which the organ resides; visceral layers wrap around organs. Synovial membrane is found in the spaces between joints and produces synovial fluid which reduces friction between joints during movement. (Note: sometimes the meninges of the brain are classified as a separate (fifth) class of membrane.)

18 **There are three major types of tissue in the human body.**

There are *four* main tissue types: *epithelial, connective, muscle* and *nervous*. Epithelial tissue covers and lines much of the body. Connective is the most common tissue, holding things together it provides structure and support. It can be loosely arranged (such as adipose tissue, collagen) or more densely packed such as tendons and ligaments. Blood and lymph are types of connective tissue. There are three types of muscle tissue: skeletal, cardiac and smooth (see Chapter 4). Nervous tissue facilitates transmission and coordination of electrical impulses throughout the body in response to various stimuli.

19 **A gland is a single cell or group of cells adapted for secretion.**

Glands are specialized epithelial cells involved in secretion. Endocrine glands secrete hormones into the circulation, for example, the thyroid gland secretes thyroxine into the blood. Exocrine glands secrete into ducts or directly onto a free surface; examples include sweat, mucous, digestive and oil glands.

20 | **There are 10 major body systems.**

The 10 organ systems are: *musculoskeletal, integumentary, nervous, endocrine, cardiovascular, respiratory, lymphatic/immune, gastrointestinal, urinary* and *reproductive.*

 MULTIPLE CHOICE

Correct answers identified in bold italics.

21 | **Cellular respiration is defined as:**

a) an intracellular, energy-producing process   b) an extracellular, energy-producing process   c) an intracellular, energy-requiring process   d) an extracellular, energy-requiring process

Cellular respiration is the process by which a cell obtains oxygen, distributes it to the mitochondria and uses it to break down sugars (such as glucose) to produce energy for the cell. The by-products of cellular respiration are carbon dioxide (CO_2) and water (H_2O).

22 | **Cellular diffusion is where solutes:**

a) move from a low concentration to a high concentration
b) move from a high concentration to a low concentration   c) stay in their relative chambers, due to other pressures within the body   d) stay in their relative chambers, unaffected by other pressures within the body

Diffusion refers to the passive movement of solvents or solutes (such as gas molecules or electrolytes). When solvents or solutes are present at a higher concentration relative to their concentration in a neighbouring chamber, the solvent or solutes will move from the area of higher concentration to the area of lower concentration until they are evenly dispersed and when no further net exchange will occur. This is known as equilibrium.

23 | **Osmosis describes the process when water:**

a) moves across a membrane from an area of higher concentration (with lower solutes) to lower concentration (with higher solutes)
b) moves across a membrane from an area of lower concentration (with higher solutes) to higher concentration (with lower solutes)   c) stays in its relative chamber, unaffected by dissolved solutes   d) stays in its relative chamber, due to the presence of dissolved solutes

Water moves, via osmosis, from an area of higher water concentration, where the solute concentration is lower, to an area of lower water concentration, where the solute concentration is higher. Equilibrium occurs where there are equal parts of water and solutes in both areas.

24 **Phagocytosis is where cells**

a) ingest and destroy microbes, cell debris and other foreign matter   b) export substances from the cell to the extracellular space   c) move fluid from the extracellular space to within the cell   d) produce erythrocytes for release within the systemic system

Phagocytosis is the process whereby phagocytic cells engulf and destroy bacteria and other foreign substances by developing projections of the plasma membrane (called pseudopods), which surround, fuse and then ingest the foreign body as it enters the cytoplasm where it is then digested by enzymes.

25 **Prokaryotes lack:**

a) a cell membrane   b) a ribosome   c) cytoplasm   *d) a cell nucleus*

A prokaryote is a bacterial organism that lacks a cell nucleus, mitochondria and other membrane bound organelles. Organisms that have membrane bound organelles are known as eukaryotes.

FILL IN THE BLANKS

26 **The science of body structure is called _anatomy_; the study of body function is _physiology_.**

Anatomy is the study of structure and organization of body parts whereas physiology is the study of the function of body parts (what they do) *and* how they do it. These terms apply to the normal, healthy state of the body. Pathophysiology (or pathobiology) describes changes to normal physiology due to disease or illness.

27 **The _mitochondria_ (sing. _mitochondrion_) are the 'powerhouses' of the cell because they produce energy in the form of ATP.**

The mitochondria convert adenosine *di*phosphate (ADP) to adenosine *tri*phosphate (ATP) by adding one phosphate molecule. ATP is a form of chemical energy that the cell uses to function. When the body needs energy (such as during muscle contraction), the ATP molecule breaks down, releasing energy.

28 **The genetic material (DNA) is found in the _nucleus_.**

Sequences of the DNA are formed into genes which are found on chromosomes in the nucleus. DNA is also replicated in the nucleus during cell division.

29 *Organelles* **are structures within cells that perform cellular functions.**

These specialized subunits within cells perform specific functions. They are usually enclosed within their own phospholipid membrane. Examples include mitochondria, lysosomes, Golgi apparatus and ribosomes.

30 **The cellular organelle responsible for digesting exhausted organelles is the *lysosome*.**

Sometimes called the suicide sac. In addition to engulfing or digesting excess (or worn-out) organelles and food particles, lysosomes have a major role in defending the cell against invading foreign bodies. Lysosomal membranes isolate these destructive enzymes to protect the rest of the cell.

31 **The study of cells is called *cytology*; the study of tissues is called *histology*.**

These methodologies usually require the use of a microscope to view cells and tissues. To accurately view most cells/tissues, samples must be carefully prepared without compromising their structure.

32 **Blood is a type of *connective* tissue.**

Although blood is a viscous fluid, it is classified as a specialized form of connective tissue since it delivers necessary substances (nutrients and oxygen) to the body's cells. Blood originates in the bone marrow and is comprised mainly of red cells (erythrocytes), white cells (leucocytes) and platelets (thrombocytes), all suspended in plasma. Other types of connective tissue include adipose tissue and cartilage. Collectively, connective tissue is the most abundant tissue in the body.

2 Essential biology and biochemistry

INTRODUCTION

Biochemistry is the study of the chemical processes in living organisms. It explores the structure and function of cellular components such as proteins, carbohydrates, lipids, nucleic acids and other biomolecules. It involves various aspects including pH, fluids and energy requirements.

The atom is made up of protons, neutrons and electrons. An atom donates an electron to become a positive ion or gains an electron to become a negative ion. Groups of atoms form molecules or compounds. A molecule consists of two or more chemically combined atoms. A compound is a pure substance consisting of two or more different chemical elements that can be separated into simpler substances by chemical reactions.

pH is a measure of acidity or alkalinity. It is measured on a scale of 1–14 with pH 7 being neutral. A pH value less than 7 is acidic; a pH greater than 7 is alkaline (or basic). The pH of bodily fluids must remain relatively constant for the body to maintain homeostasis.

Water is the most abundant substance in the body. Adenosine triphosphate (ATP) is the main energy storage molecule in humans. It is important for paramedics to know about the essential nutrients that the body needs and how it uses them, because lack of nutrients or a disruption of the body's chemistry can lead to illness or disease.

Useful resources

Mader's Understanding Human Anatomy and Physiology
Chapter 2

Anatomy and Physiology (8th edition)
Chapters 1, 2, 3 and 4

BBC Chemistry:
http://www.bbc.co.uk/schools/gcsebitesize/chemistry/

Animation of the sodium-potassium pump:
http://highered.mcgraw-hill.com/sites/0072495855/student_view0/chapter2/
animation__how_the_sodium_potassium_pump_works

LABELLING EXERCISE

1–10 Use the words in the box to complete the pH scale in Figure 2.1. In numbers 1–3 identify the trends on the scale, in numbers 4–10 identify the body fluids with the specified pH values.
Not all of them are required, so choose carefully!

semen	tears
blood	urine
neutral	stomach acid
mucus	cerebrospinal fluid (CSF)
saliva	acidic
alkaline (basic)	bile
pancreatic juice	

Figure 2.1 The pH scale

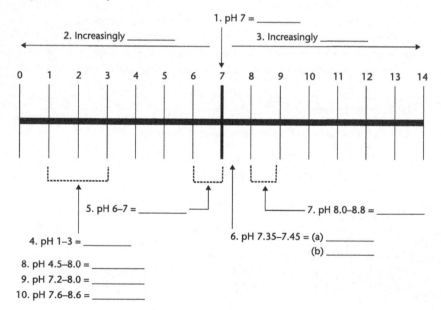

1. pH 7 = _____
2. Increasingly _____
3. Increasingly _____
5. pH 6–7 = _____
7. pH 8.0–8.8 = _____
4. pH 1–3 = _____
6. pH 7.35–7.45 = (a) _____
(b) _____
8. pH 4.5–8.0 = _____
9. pH 7.2–8.0 = _____
10. pH 7.6–8.6 = _____

TRUE OR FALSE?

Are the following statements true or false?

11 Adenosine triphosphate temporarily stores and transfers energy to cellular activities that require energy.

12 An anabolic reaction is a synthesis reaction.

13 Carbohydrates contain carbon, nitrogen and oxygen.

14 Lipids contain carbon, nitrogen and oxygen.

15 Proteins are made up of amino acids.

16 Active transport is a passive process.

17 Water can pass through a selectively permeable membrane.

MULTIPLE CHOICE

Identify one correct answer for each of the following.

18 An enzyme is described as:

a) a carbohydrate that slows down reactions
b) a protein that slows down reactions
c) a carbohydrate that speeds up reactions
d) a protein that speeds up reactions

19 In the human body, enzyme activity is most efficient at:

a) 0°C
b) 37°C
c) 42°C
d) 60°C

20 Deoxyribonucleic acid is:

a) double-stranded and anti-parallel
b) single-stranded and anti-parallel
c) double-stranded and parallel
d) single-stranded and parallel

21 A positively charged electrolyte is called:

a) a cation
b) an anion
c) an isotope
d) a positron

22　A pH of 7.1 is considered:

　　a)　neutral
　　b)　weakly acidic
　　c)　weakly alkaline
　　d)　strongly acidic

23　Which of these is not used by the body for energy?

　　a)　fat
　　b)　carbohydrate
　　c)　protein
　　d)　water

24　Which dietary component does the body most commonly use for energy?

　　a)　fats
　　b)　carbohydrates
　　c)　proteins
　　d)　vitamins

25　Monosaccharides are:

　　a)　proteins
　　b)　fatty acids
　　c)　simple sugars
　　d)　complex carbohydrates

FILL IN THE BLANKS

Fill in the blanks in each statement using the options in the box below.
Not all of them are required, so choose carefully!

haemoglobin	hypertonic
catabolism	insoluble
anabolism	diffusion
osmosis	isotonic
soluble	active transport
metabolism	antibody

26 An _____ is a protein in the immune system.

27 _____ is a protein with a quaternary structure.

28 Lipids are _____ in water.

29 The sum of all the body's chemical processes is its _____.
This consists of two parts: the phase that builds up new substances
is called _____ and the phase that breaks substances down is
called _____.

30 If the concentrations of solutes in the extracellular fluid and intracellular
fluid are equal, the cell is in an _____ solution.

31 _____ is the movement of water from an area of high water concentration to an area of lower water concentration across a selectively
permeable membrane.

ANSWERS

LABELLING EXERCISE

Figure 2.2 The pH scale

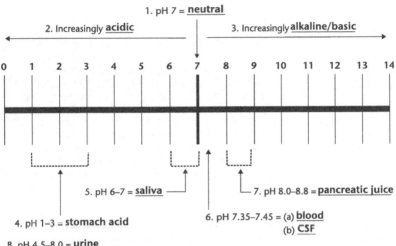

1. pH 7 = <u>neutral</u>
2. Increasingly <u>acidic</u>
3. Increasingly <u>alkaline/basic</u>

0 1 2 3 4 5 6 7 8 9 10 11 12 13 14

5. pH 6–7 = <u>saliva</u>
7. pH 8.0–8.8 = <u>pancreatic juice</u>
4. pH 1–3 = <u>stomach acid</u>
6. pH 7.35–7.45 = (a) <u>blood</u>
      (b) <u>CSF</u>
8. pH 4.5–8.0 = <u>urine</u>
9. pH 7.2–8.0 = <u>semen</u>
10. pH 7.6–8.6 = <u>bile</u>

1   *Neutral:* on the pH scale, 7.0 is considered neutral, it is neither acidic nor alkaline (basic). pH stands for 'power of <u>h</u>ydrogen' and the scale measures the concentration of hydrogen ions and hence acidity or alkalinity of a solution. The scale is numbered 0–14. By convention, 'p' is always a lower case letter, 'H' is always a capital letter and the letters 'pH' are always before the number (that is, it is always pH 7.0 or pH 4.5 and not 7 pH). The strength of the acid or alkali changes by a factor of ten between each pH value, therefore an acid with pH 4 is ten times stronger than an acid of pH 5. It is difficult to eliminate acidity or alkalinity simply by diluting – it must be neutralized by adding an acid to alkali or vice versa.

2   *Increasingly acidic:* any pH less than 7.0 is acidic and the further away from 7.0, the stronger the acid, that is, an acid with pH 2 (such as lemon juice) is much stronger than an acid with pH 4.5 (such as tomato juice).

3 *Increasingly alkaline/basic:* any pH greater than 7.0 is alkaline (or basic) and the further away from 7.0, the stronger the alkali. An alkali of pH 13 (such as oven cleaner) is much stronger than an alkali with pH 10 (such as antacid tablets).

4 *Stomach acid:* pH 1–3, a highly acidic solution in the stomach. It contains a lot of hydrochloric acid (HCl) which is very acidic. HCl acid is secreted by the parietal cells in the stomach and is vital for breaking down food during digestion.

5 *Saliva:* the pH of saliva is slightly acidic, normally around pH 6–7. It is generally a clear secretion formed mainly in the parotid, submandibular and sublingual glands. It has lubricating, cleansing, anti-microbial, excretory and digestive functions.

6 *(a) Blood:* the pH of the blood is slightly alkaline. It is tightly maintained within the range pH 7.35–7.45. Enzyme activity in the body may be affected if the blood pH falls outside this range. In the body, the blood is a buffer solution. This means that when the body is functioning normally, the blood can counteract very small changes in pH. In certain diseases, the blood pH can drift outside these limits and the body will then attempt to bring it back to normal. If the blood pH deviates too far outside this range, many vital functions, such as renal filtration and respiration, may be adversely affected. If the blood pH is not corrected, a patient may eventually fall into a coma.

6 *(b) Cerebrospinal fluid (CSF):* has a very similar pH to blood. It is usually a clear fluid produced by the choroid plexus to protect the brain from infection and trauma.

7 *Pancreatic juice:* a pH between 8 and 8.8 makes pancreatic secretions one of the most alkaline body fluids. This is a clear, alkaline secretion of the pancreas secreted through the pancreatic duct into the duodenum and contains enzymes that aid digestion of proteins, carbohydrates and fats. The high concentration of bicarbonate (also known as hydrogencarbonate, HCO_3^-) ions helps to neutralize acidic gastric secretions produced by the stomach.

8 *Urine:* the liquid waste secreted by kidneys during urination (micturition) via the urethra. The pH of urine can vary quite dramatically compared to other bodily fluids, depending on diet and metabolic state. It can range between pH 4.5 and 8.0 but in Western populations it is usually acidic (around pH 6.0) due to diets high in protein. Urinary pH fluctuates according to the time of day and in normal physiology it is usually slightly more alkaline upon wakening, moving closer to neutral pH as the day progresses. Acidic urine can contribute to the formation of 'stones' (renal calculi) in the kidneys, ureters or bladder.

9   **Semen:** a mixture of sperm and seminal fluid, it usually has a pH 7.2–8.0. The seminal fluid provides a transport medium and nutrients for sperm.

10   **Bile:** with a pH 7.6–8.6, bile is a slightly alkaline fluid. Usually brownish-yellow or greenish-yellow, bile is secreted by the liver and concentrated in the gall bladder. It enters the small intestine via the bile duct. After a meal (especially a high-fat meal), the gall bladder contracts and discharges bile. Bile helps digestion by emulsifying fats and maintaining the alkalinity of the intestinal contents. The main constituents of bile are conjugated bile salts, cholesterol, phospholipid, bilirubin and electrolytes.

TRUE OR FALSE?

11   **Adenosine triphosphate temporarily stores and transfers energy to cellular activities that require energy.**

Adenosine triphosphate (ATP) is responsible for intracellular energy transfer, transporting chemical energy within cells for metabolism. Generally, the cellular activities that ATP enables are muscular contraction, movement of chromosomes during cell division and transporting substances across cell membranes.

12   **An anabolic reaction is a synthesis reaction.**

In an anabolic reaction, two or more atoms, ions or molecules combine to form a new, larger molecule. The opposite of this is a catabolic reaction. In this decomposition reaction a molecule splits into smaller molecules, atoms or ions.

13   **Carbohydrates contain carbon, nitrogen and oxygen.**

Carbohydrates are simple organic molecules that contain only carbon, *hydrogen* and oxygen. They are the most abundant of the four major classes of biomolecules. They have vital roles in the storage and transport of energy in the body. Most carbohydrates are digested into simple sugars which can then be absorbed into the blood for use by the body. The smallest unit of a carbohydrate is a monosaccharide, such as glucose, fructose or galactose. These come together in anabolic reactions, forming disaccharides through the loss of a water molecule in a dehydration reaction. Disaccharides include sucrose (glucose and fructose) and lactose (glucose and galactose). When many disaccharides bond together, they form complex polysaccharides, such as starch (mainly in plants) or glycogen (in humans). Carbohydrates are the main source of energy for the body since they are the easiest nutrient to digest and absorb and most readily used as energy in the organs and tissues.

14 **Lipids contain carbon, nitrogen and oxygen.**

Lipid (fat) molecules (as with carbohydrates) contain carbon, *hydrogen* and oxygen. They differ from carbohydrates because the ratio of the carbon:hydrogen:oxygen elements in their molecules varies. There are four major classes of lipids: phospholipids, glyercides, sterols and prostaglandins. Lipids are insoluble in water and often contain phosphorous elements – as in the phospholipids that make up many selectively permeable biological membranes.

15 **Proteins are made up of amino acids.**

The smallest unit of a protein is an amino acid. Amino acids combine in anabolic reactions to form dipeptides (two amino acids), tripeptides (three amino acids) and polypeptides (many amino acids). Proteins (or polypeptides) are large organic molecules that always contain carbon, hydrogen, oxygen and nitrogen; some proteins also contain sulphur. Proteins are more complex than carbohydrates or lipids and their functions are determined by their structure.

16 **Active transport is a passive process.**

Active transport requires energy, therefore it is not a passive process. Energy is required because it involves moving a substance across a cell membrane but *against* a concentration gradient, that is, from an area of low concentration to an area of higher concentration. Special proteins within the cell membrane act as 'carrier proteins' to facilitate active transport. The energy is provided by adenosine triphosphate (ATP) which is generated in the mitochondria. Examples of biological processes that use active transport include the sodium-potassium pump in nerve cells and re-absorption of glucose, amino acids and salts through the proximal convoluted tubule of the nephron in the kidney.

17 **Water can pass through a selectively permeable membrane.**

A selectively permeable membrane is a biological membrane made up of a double layer of phospholipid molecules. It allows water to pass through but certain substances (solutes, such as sodium ions or blood plasma proteins) cannot pass through due to their relatively large size. Many cell membranes in the body are selectively permeable; this determines the movement of solutions and solutes across membranes.

a b c d   MULTIPLE CHOICE

Correct answers identified in bold italics.

18 **An enzyme is described as:**

a) a carbohydrate that slows down reactions   b) a protein that slows down reactions   c) a carbohydrate that speeds up reactions *d) a protein that speeds up reactions*

Enzymes are 'biological catalysts' that speed up the rate of reactions in the body. They are highly effective because without enzymes many biological reactions would not occur fast enough to sustain life. Enzymes are highly specific; each enzyme will only work on a specific substance (substrate) or closely related family of substrates. The substrate can often be identified by the enzyme name: since the prefix usually takes the name of the substrate and the suffix '-ase' indicates an enzyme. For example, the enzyme sucrase helps digest the disaccharide sucrose into its constituent monosaccharides, glucose and fructose. Enzymes have three important properties: (1) specificity for their substrate; (2) efficiency; and (3) control.

19 **In the human body, enzyme activity is most efficient at:**

a) 0°C   *b) 37°C*   c) 42°C   d) 60°C

This is also known as 'core body temperature'. This refers to the temperature of organs in the body; the temperature at the body surface (skin) is called peripheral body temperature. This can differ significantly from core body temperature, for example, the temperature of the toes can be as low as 29°C. This difference is due to changes in the temperature of the blood as it flows around the body. It is important that the core body temperature is maintained within the acceptable range since many body functions rely on enzymes, which are all temperature-sensitive.

20 **Deoxyribonucleic acid is:**

a) double-stranded and anti-parallel   b) single-stranded and anti-parallel   c) double-stranded and parallel   d) single-stranded and parallel

Deoxyribonucleic acid (DNA) carries the cell's genetic information and is the 'blueprint' for protein synthesis. DNA molecules exist as pairs (double-stranded chain), forming a spiral shape called the double helix. These two strands run in opposite directions to each other and are therefore anti-parallel.

21 **A positively charged electrolyte is called:**

a) a cation   b) an anion   c) an isotope   d) a positron

An ion is an atom or molecule in which the total number of electrons is not equal to the total number of protons, giving it an overall positive or negative electrical charge. Electrons are negatively charged and protons are positively charged. An anion is an ion with more electrons than protons, giving it an overall negative charge, for example, chloride ion, Cl^-. A cation is an ion with more protons than electrons, giving it an overall net positive charge, for example, sodium ion, Na^+; magnesium ion, Mg^{2+}. *Hint: to remember that cations are positive, think 't' (in cation) looks similar to a '+' sign.*

---

22 **A pH of 7.1 is considered:**

a) neutral   b) weakly acidic   *c) weakly alkaline*   d) strongly acidic

Any substance with a pH greater than 7.0 is alkaline (or basic) and alkalinity increases as the number gets further away from 7.0. Substances with pH less than 7.0 are acidic and acidity increases as the pH number moves further away from 7.0. Maintaining the correct pH in the human body is very important because many of its physiological functions are dependent on accurate pH to maintain enzyme function.

23 **Which of these is not used by the body for energy?**

a) fat   b) carbohydrate   c) protein   *d) water*

Pure water contains no calories and therefore does not provide the body with energy. The human body is anywhere from 50–80 per cent water depending on gender, age and body size. At birth, a baby's body weight may be as much as 75–80 per cent water. As the newborn loses water during the first few days of life, the body weight usually drops slightly. In the older adult, the volume of bodily fluids decreases so the body retains less water. Females generally have less water than males since they have more adipose (fat) tissue.

24 **Which dietary component does the body most commonly use for energy?**

a) fats   *b) carbohydrates*   c) proteins   d) vitamins

There are seven components of the diet: carbohydrates, lipids, proteins, vitamins, minerals, fibre and water. Not all of these components provide the body with energy. Food provides energy when its nutrients are broken down (digested). The main source of energy for the body is the monosaccharide, glucose. The production of the energy molecule adenosine triphosphate (ATP) requires the breakdown of glucose. Fats and proteins will also provide energy but only in the absence of available glucose (for example, diabetes).

25 **Monosaccharides are:**

a) proteins   b) fatty acids   *c) simple sugars*   d) complex carbohydrates

Simple sugars (monosaccharides) are the smallest form of carbohydrate. The common monosaccharides are glucose, fructose and galactose. Complex carbohydrates (starch or glycogen) are longer chains of carbohydrates (also called oligosaccharides or polysaccharides) that require breakdown, by appropriate enzymes, into their simplest form. Complex carbohydrates provide a slower, more sustained release of energy than simple carbohydrates and so keep blood-sugar (glucose) levels stable for longer; this is particularly important for diabetic patients.

FILL IN THE BLANKS

26 An *antibody* is a protein in the immune system.

Antibodies are plasma proteins sometimes called immunoglobulins (abbreviated as Ig). They are also classified as glycoproteins, which are proteins attached to a sugar molecule. They are found in the blood and used by the immune system to identify and neutralize foreign invaders, such as bacteria and viruses.

27 *Haemoglobin* is a protein with a quaternary structure.

Quaternary proteins are the most complex types of protein, for example, the iron-containing protein haemoglobin. The primary structure of a protein is the sequence of amino acids in the protein chain. A secondary protein structure results from bonding between parts of the primary chain. This can produce different shapes, the most common being the alpha-helix, another being the beta-pleated sheet. If a secondary structure folds over on itself, it forms a tertiary protein structure. When a number of tertiary proteins are combined, this forms a quaternary structure.

28 Lipids are *insoluble* in water.

Lipids are hydrophobic (or water-repelling) in nature. They are a large family group of naturally-occurring molecules which includes fats, waxes, sterols, fat-soluble vitamins (such as vitamins A, D, E and K), monoglycerides, diglycerides, phospholipids, among others. The main biological functions of lipids include energy storage, structural components of cell membrane and important biochemical signalling.

29 The sum of all the body's chemical processes is its *metabolism*. This consists of two parts: the phase that builds up new substances is called *anabolism* and the phase that breaks substances down is called *catabolism*.

The process of catabolism extracts energy from substances, in the form of ATP, that can be used by cells.

30 If the concentrations of solutes in the extracellular fluid and intracellular fluid are equal, the cell is in an *isotonic* solution.

Isotonic solutions are in equilibrium, meaning there is no net (overall) movement of water in or out of the environment, such as a cell. The two extremes of isotonic are hypertonic and hypotonic. In a hypertonic environment, cells are in a highly concentrated solution, meaning there is more water in the cells and so water leaves cells and moves out into the environment, aiming to reach equilibrium. In a hypertonic solution, cells will shrink because they lose water. In a hypotonic environment there is a high concentration of water, meaning that cells contain less water than their environment so water diffuses out of the environment into the cells, causing cells to swell. Because of the increased volume of water inside them, cells may eventually burst if they swell too much (see Figure 2.3).

Figure 2.3 Comparison of water movement in and out of cells

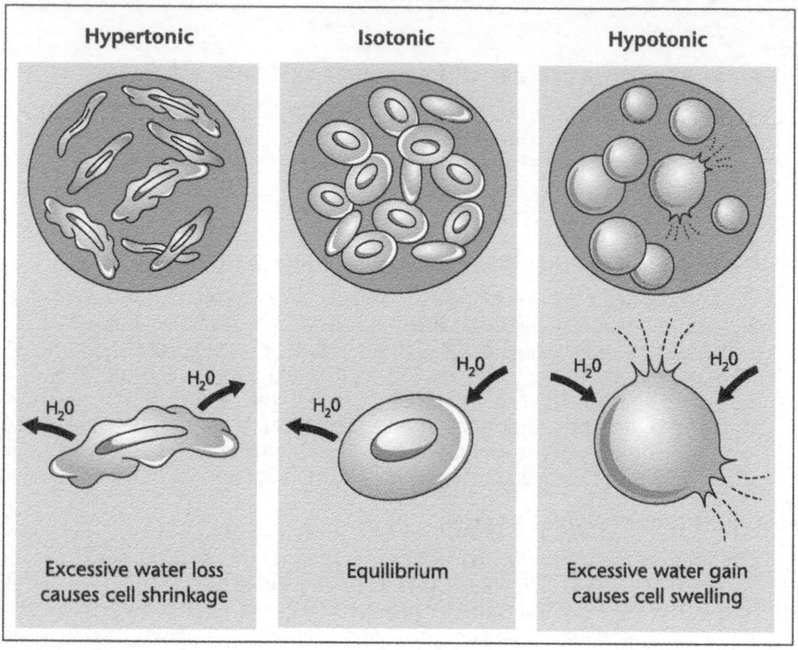

|Hypertonic|Isotonic|Hypotonic|

Excessive water loss causes cell shrinkage | Equilibrium | Excessive water gain causes cell swelling

31   *Osmosis* **is the movement of water from an area of high water concentration to an area of lower water concentration across a selectively permeable membrane.**

Osmosis specifically refers to the diffusion of water. Diffusion refers to the movement of solvents and solutes from an area of high concentration to an area of lower concentration down a concentration gradient. Diffusion and osmosis are passive processes – they do not require energy (think about free-wheeling a bicycle down a hill – no pedalling is required because of the 'downhill' gradient). When diffusion or osmosis occurs against a concentration gradient (that is, from lower concentration to higher concentration), it requires energy in the form of ATP and is called active transport (think about cycling up a hill – a lot of pedalling is needed because of the 'uphill' gradient and pedalling requires energy!).

An alternative osmosis definition referring to solute concentration states osmosis is the movement of water from an area of low *solute* concentration to higher *solute* concentration across a selectively permeable membrane.

Figure 2.4 Osmosis

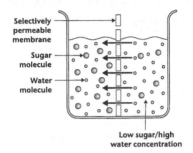

Selectively permeable membrane

Sugar molecule

Water molecule

Low sugar/high water concentration

. Figure 2.4 illustrates osmosis – in the concentrated sugar solution (left side) there is less water while in the dilute sugar solution (right side) there is more water. This water will move through the selectively permeable membrane from the region of high water content, down the concentration gradient, to the region where water is less concentrated because more solutes are dissolved in it. This movement of water will cause an increase in water volume on the left side of the container and reduced volume on the right.

3 The integumentary system

INTRODUCTION

The skin (integument) is the outer covering of the body comprising different tissues that perform specific functions. These include the maintenance and regulation of body temperature, protection, sensation, excretion and immunity.

Many factors affect both the appearance and health of skin, including nutrition, hygiene, circulation, age, genetic traits and immune status. Environmental stress, psychological trauma and drug use may also alter its appearance. Furthermore, because of its visibility, the skin often reflects our state of emotion, along with aspects of normal physiology.

The skin is the largest organ of the body and examination of its condition can often prove invaluable for medical diagnosis. Its exterior location makes it vulnerable to damage from trauma, UV light, microbes and environmental chemicals. Severely damaged skin will try to heal by forming scar tissue.

Many pre-hospital interventions will involve the skin, so paramedics should know and understand how the skin protects the body, helps maintain homeostasis and encloses the organ systems.

Useful resources

Paramedics! Test Yourself in Pathophysiology

Chapter 3

Ross and Wilson's Anatomy and Physiology in Health and Illness (11th edition)

Chapter 14

Anatomy and physiology of the skin:

http://www.virtualmedicalcentre.com/anatomy.asp?sid=2

More information on dermatology, the skin and skin conditions:

http://dermatology.about.com/od/dermatologybasics/Dermatology_Basics.htm

LABELLING EXERCISE

1–10 Identify the regions and structures of the skin and hair shaft in Figure 3.1, using the options provided in the box below.

hair shaft	subcutaneous tissue
adipose tissue	sebaceous gland
arrector pili muscle	hair follicle
sweat gland	dermis
epidermis	hair bulb

Figure 3.1 The skin and hair shaft

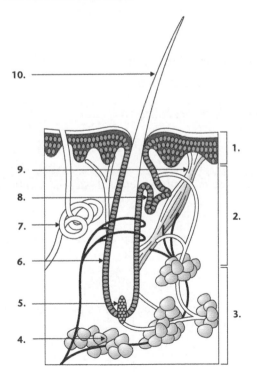

TRUE OR FALSE?

Are the following statements true or false?

11 The dermis acts as a blood reservoir.

12 Sweat removes certain organic compounds from the body.

13 The subcutaneous layer consists of areolar and adipose tissue.

14 Melanocytes are found in the dermis.

15 The dermis is composed of five basic strata.

16 The epidermis contains specialized receptors involved in the sensation of touch.

17 The process of epidermal cell development is called keratinization.

18 Certain cells of the epidermis are important components of the immune system.

19 The elasticity of skin is mainly due to the adipose tissue in the dermis.

20 Pacinian corpuscles in the subcutaneous layer are temperature-sensitive.

21 Deep wound healing has two main stages.

22 Thermoregulation by the skin involves cooperation between numerous positive feedback mechanisms.

23 There are three types of glands associated with the skin.

MULTIPLE CHOICE

Identify one correct answer for each of the following.

24 There are three layers of epithelium:

a) transitional, intertransitional, and reinforced

b) squamous, cubodial and columnar

c) simple, stratified and pseudostratified

d) elastic, reticular and stroma

25 Skin colour is determined by which pigment?

a) melanin

b) carotene

c) haemoglobin

d) all of the above

26 The sebaceous glands in the skin produce:

a) milk

b) oil

c) sweat

d) cerumen

27 Which of the following structures is not an accessory component of the skin?

a) nail

b) subcutaneous tissue

c) hair

d) mammary glands

28 Which tissue region do nails originate from?

a) eponychium

b) lunula

c) nail matrix

d) cuticle

29 The skin fibres are arranged in bundles known as:

a) zone of hyperaemia

b) lines of cleavage

c) epidermal ridges

d) stratum lucidum

30 The waterproof coating found in epidermal cells is called:

a) myelin

b) keratin

c) melanin

d) albumin

31 Which of the following can pass most easily through the epidermis?

a) proteins

b) lipid-soluble molecules

c) water-soluble compounds

d) salts

32 Which of the following substances is not present in sweat?

a) urea

b) calcium

c) lactic acid

d) water

33 The dermis contains which of the following?

a) blood vessels

b) sweat glands

c) sensory nerve endings

d) sebaceous glands

MATCH THE TERMS

Identify which statement matches each description below.

A. Papillary layer of the dermis   **E.** Epidermal growth factor
B. Dermis   **F.** Stratum corneum
C. Stratum lucidum   **G.** Epidermis
D. Reticular layer of the dermis   **H.** Stratum basale

34 Translucent cells, containing keratin ____

35 Layer of dead cells ____

36 Deep region of the dermis ____

37 Origin of many accessory structures ____

38 Site of Meissner's corpuscles (type of nerve ending) ____

39 Principal superficial region of skin ____

40 A protein that stimulates growth of cells during tissue repair and renewal ____

41 Region involved in rapid cell division ____

ANSWERS

LABELLING EXERCISE

Figure 3.2 The skin and hair shaft

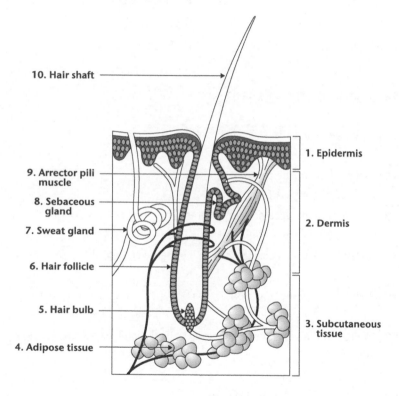

1. **Epidermis:** the superficial part of the skin composed of keratinized squamous epithelial tissue. Four or five distinct cell layers form the epidermis. In most regions of the body the epidermis is approximately 0.1 mm thick and has four layers. Areas subjected to most friction, such as the palms and soles, contain five layers that are typically thicker (1–2 mm). Approximately 90 per cent of epidermal cells are keratinocytes that produce the protein keratin which helps waterproof and protect the skin from its external environment. The epidermis is connected to the deeper, thicker connective tissue that comprises the dermis.

2   **Dermis:** the second main part of the skin; composed of connective tissue containing collagen and elastic fibres. The dermis is relatively thick in the palms and soles and is much thinner in areas such as the eyelids and scrotum. Blood vessels, nerves, glands and hair follicles are all embedded in the dermis.

3   **Subcutaneous tissue:** provides functional support for the dermis and epidermis. It also contains the major blood vessels that supply the skin and underlying adipose tissue. The subcutaneous layer also contains the Pacinian corpuscles that are sensitive to pressure and numerous nerve endings sensitive to temperature.

4   **Adipose tissue:** in the subcutaneous layer of skin this provides insulation and conserves body heat, in addition to preventing heat gain from the external environment.

5   **Hair bulb:** found at the base of each hair follicle, it contains the hair papilla which contains blood vessels that supply the growing hair. The bulb also contains a ring of matrix cells that stimulate growth of existing hairs. New hairs are produced through cell division when older hairs are lost through shedding. This replacement occurs within the same follicle.

6   **Hair follicle:** a structure composed of an internal and external root sheath. The external sheath is a downward continuation of the epidermis. The internal sheath forms a tubular structure that lines the hair follicle.

7   **Sweat gland:** (exocrine) glands are widely distributed in the skin. These glands are divided into two main types: eccrine and apocrine glands. Eccrine glands regulate body temperature through the production of sweat, while apocrine glands produce sweat during periods of emotional upset, fright or pain.

8   **Sebaceous gland:** ductless glands that open directly onto the surface skin at specific locations or more commonly into the hair follicles. Absent in the palms and soles, sebaceous glands vary in size and shape depending on their location.

These glands produce a mixture of oil and sebum (cell debris), which prevents excess water evaporation from the skin. It also helps maintain the soft and pliable texture of the skin and hair and acts to inhibit the growth of certain bacteria.

9   **Arrector pili muscle:** a smooth muscle that contracts under stress and when the body is exposed to cold, pulling hairs into a vertical position (piloerection).

10   **Hair shaft:** comprised of columns of dead keratinized cells derived from actively growing epidermal cells located within the hair follicle. As these epidermal cells grow and divide, older cells are pushed towards the surface. The shaft is the superficial portion of the hair which projects from the surface of the skin, and hair is distributed throughout the body. The main function of hair is protection of sensitive structures of the body.

TRUE OR FALSE?

11 **The dermis acts as a blood reservoir.**

The dermis contains an extensive network of blood vessels that carry approximately 15 per cent of the total blood flow in a resting adult. During moderate exercise the blood vessels dilate, increasing blood flow and hence promoting heat loss through the skin surface.

12 **Sweat removes certain organic compounds from the body.**

Besides removing heat and some water from the body, sweat also allows for the loss of a small quantity of ions and several organic compounds such as urea and lactic acid.

13 **The subcutaneous layer consists of areolar and adipose tissue.**

The subcutaneous layer of skin consists of loose connective (areolar) and fat (adipose) tissues. The collagenous and elastic fibres of this layer are continuous with those of the dermis and run parallel to the surface of the skin, extending in all directions. As a result of this layout, no definitive boundary exists between the dermis and subcutaneous layers.

14 **Melanocytes are found in the dermis.**

Melanocytes produce the pigment melanin and are found in the stratum basale of the epidermis. They possess slender projections called dendrites that transfer melanin to keratinocytes.

15 **The dermis is composed of five basic strata.**

The dermis is composed of two layers: the superficial papillary region and a deeper reticular region. The papillary region is composed of areolar connective tissue and the reticular region is composed of dense, irregular connective tissue.

16 **The epidermis contains specialized receptors involved in the sensation of touch.**

These receptors, called Merkel cells, are attached to keratinocytes in the stratum basale of hairless skin and are involved in touch sensation. Merkel cells make contact with the flattened end of sensory neurones and initiate the sensation of touch.

17   **The process of epidermal cell development is called keratinization.**

During keratinization, cells formed in the stratum basale undergo a developmental and maturation process as they are pushed towards the surface. During this process they accumulate keratin while losing cytoplasm and cellular organelles. These keratinized cells eventually slough off the skin and are replaced by underlying cells which also undergo keratinization. This process, from cell formation deep in the epidermis until removal by sloughing, takes approximately 28 days.

18   **Certain cells of the epidermis are important components of the immune system.**

Langerhans cells are an important population of cells which arise in bone marrow and migrate to the epidermis where they interact with T-helper cells of the immune system.

19   **The elasticity of skin is mainly due to adipose tissue in the dermis.**

The combination of collagen and elastic fibres deep in the dermis provide the skin with strength and elasticity. The stretching ability of skin is demonstrated in pregnancy, obesity or as a result of oedema. Tears that occur during extreme stretching and remain visible afterwards are termed striae (stretch marks), they are often observed in previously obese individuals and are common in pregnancy.

20   **Pacinian corpuscles in the subcutaneous layer are temperature-sensitive.**

Pacinian corpuscles are sensitive to pressure. Nerve endings sensitive to cold are located in the dermis, while those sensitive to heat are found in the intermediate and superficial dermis.

21   **Deep wound healing has two main stages.**

Deep wounds describe skin injuries that extend into the dermis. The repair process is complex and often results in scar formation. It involves four stages: inflammation, migration, proliferation and maturation.

22   **Thermoregulation by the skin involves cooperation between numerous positive feedback mechanisms.**

Negative feedback mechanisms ensure that core and surface body temperature remain relatively constant. The skin plays a significant role in the homeostatic regulation of body temperature. Despite large fluctuations in environmental temperature, the negative feedback loop maintains body temperature within acceptable limits.

23 | There are three types of glands associated with the skin.

Sebaceous glands are mainly connected to hair follicles and secrete sebum, which keeps hair from drying, reduces the evaporation of water from the skin and stops the growth of some bacteria. Ceruminous glands produce wax (cerumen) in the ear. Sudoriferous glands (sweat glands) produce sweat (mainly water and dissolved chlorides) through eccrine glands on the skin and apocrine glands, found mainly in the axilla and pubic regions. Another type of sudoriferous gland is the mammary gland, which secretes milk in females.

MULTIPLE CHOICE

Correct answer identified in bold italics.

24 **There are three layers of epithelium:**

a) transitional, intertransitional, and reinforced b) squamous, cubodial and columnar *c) simple, stratified and pseudostratified* d) elastic, reticular and stroma

Epithelium forms the superficial layer of the skin, some internal organs, the inner lining of blood vessels, ducts, and the interior of the respiratory, digestive, urinary and reproductive systems. Simple epithelium is a single layer of cells where diffusion, osmosis, filtration and secretion occur. Stratified epithelium is reinforced epithelium, where there are specific areas of 'wear and tear'. Pseudostratified epithelium has the appearance of more than one layer, as the nuclei lie at differing levels.

25 **Skin colour is determined by which pigment?**

a) melanin b) carotene c) haemoglobin *d) all of the above*

Skin colour is derived from a combination of melanin, carotene and haemoglobin. The amount of melanin in the skin varies in colour from a pale yellow to black. Melanin is located predominantly in the epidermis. Carotene, an orange pigment and a precursor of vitamin A, is found in the stratum corneum, adipose tissue, dermis and subcutaneous layers. Haemoglobin influences skin colouring through the volume of blood moving through the capillaries in the dermis.

26 **The sebaceous glands in the skin produce:**

a) milk *b) oil* c) sweat d) cerumen

The sebaceous glands produce the oils necessary to maintain skin and hair. Sweat is produced by apocrine and eccrine glands. Ceruminous glands are modified glands that produce a waxy substance which, when combined with sebaceous gland secretions, forms a sticky barrier that helps protect the external auditory canal from foreign bodies. The mammary glands produce and secrete milk.

27 **Which of the following structures is not an accessory component of the skin?**

a) nail *b) subcutaneous tissue* c) hair d) mammary glands

Subcutaneous tissue connects underlying bone and muscle with the dermis. Nail, hair and mammary glands are considered accessory skin components.

28 **Which tissue region do nails originate from?**

a) eponychium   b) lunula   *c) nail matrix*   d) cuticle

Nails are composed of tightly packed epidermal keratinocytes which grow from the nail bed (matrix). The lunula is the visible part of the nail (half-moon). It appears white because the vascular tissue underneath is not visible.

29 **The skin fibres are arranged in bundles known as:**

a) zone of hyperaemia   *b) lines of cleavage*   c) epidermal ridges
d) stratum lucidum

The lines of cleavage indicate the direction of collagen fibre bundles in the dermis and are considered during surgery.

Epidermal ridges form during foetal development and are found on the superficial surface of the palms, fingers, soles and toes. These genetically determined ridges are unique to the individual (fingerprinting).

Zones of hyperaemia relate to thermal burns and describe areas of skin that have been least damaged on exposure to heat. They are characterized by an increase in blood flow in the surrounding area.

The stratum lucidum is a layer of cells apparent in some types of epidermis, particularly the palms and soles.

30 **The waterproof coating found in epidermal cells is called:**

a) myelin   *b) keratin*   c) melanin   d) albumin

Keratinocytes within the epidermis produce the waterproof protein keratin. Melanin is the dark pigment produced by epidermal melanocytes. Myelin is a lipoprotein which is formed around the axons. Albumin is the smallest and most abundant of the plasma proteins, it helps maintain plasma colloid osmotic pressure.

31 **Which of the following can pass most easily through the epidermis?**

a) proteins   *b) lipid-soluble molecules*   c) water-soluble compounds
d) salts

Since the permeability barrier surrounding epidermal cells is composed mainly of lipids, substances which are lipid-soluble will easily diffuse though the barrier. This property is exploited in the administration of some medicines. Water-soluble substances have more difficulty diffusing through the skin because the lipid barrier prevents water loss from the body.

32 **Which of the following substances is not present in sweat?**

a) urea   *b) calcium*   c) lactic acid   d) water

Calcium is not a component of sweat but is excreted from the body in urine and faeces. Urea, lactic acid and water can all be found in sweat.

33 **The dermis contains which of the following?**

a) blood vessels   b) sweat glands   c) sensory nerve endings
d) *sebaceous glands*

Sebaceous (oil) glands originate in the dermis and are usually connected to hair follicles. Blood vessels and sensory nerve endings are typically found in the epidermis. Parts of the secretory glands are in the dermis but their secretory ducts open into the epidermis.

 MATCH THE TERMS

34 Translucent cells, containing keratin   **C.** Stratum lucidum

35 Layer of dead cells   **F.** Stratum corneum

36 Deep region of the dermis   **D.** Reticular layer of the dermis

37 Origin of many accessory structures   **B.** Dermis

38 Site of Meissner's corpuscles (type of nerve ending)   **A.** Papillary layer of the dermis

39 Principal superficial region of skin   **G.** Epidermis

40 A protein that stimulates growth of cells during tissue repair and renewal   **E.** Epidermal growth factor

41 Region involved in rapid cell division   **H.** Stratum basale

4 The musculoskeletal system

INTRODUCTION

The musculoskeletal system facilitates the movement and posture of the body and its parts. It is also known as the locomotor system and has skeletal and muscular components.

The skeletal portion provides a framework that protects the internal organs and allows coordinated movement. Bone is associated with a variety of active tissues including cartilage, dense connective tissue, blood and nervous tissue. In addition to its function in support and movement, bone is the storage system for calcium and phosphorus and contains critical components of the haematopoietic system.

Although bones provide leverage for body movement, they also require alternating contraction and relaxation of muscles across joints to provide coordinated movement.

Paramedics should appreciate the essential role of the musculoskeletal system in controlling the body's internal and external movements. It is important to remember how the anatomy and physiology of the musculoskeletal system are intimately related to the nervous system, which initiates muscle contraction.

Useful resources

Paramedics! Test Yourself in Pathophysiology
Chapter 4

Anatomy and Physiology (8th edition)
Chapters 6, 7 8 and 9

Interactive tutorial on the musculoskeletal system:
http://www.getbodysmart.com/ap/skeletalsystem/skeleton/introduction/tutorial.html

Anatomy of the musculoskeletal system:
http://www.innerbody.com/image/musfov.html

LABELLING EXERCISE

1–16   Identify the regions and structures of bone and muscle in Figure 4.1, using the options provided in the box below.

diaphysis	muscle fibre
epimysium	fascicle
articular cartilage	compact bone
medullary cavity	epiphyseal plate
nutrient artery	epiphysis
myofibril	spongy bone
deep fascia	endosteum
periosteum	perimysium

Figure 4.1 Structures and regions of bone and muscle

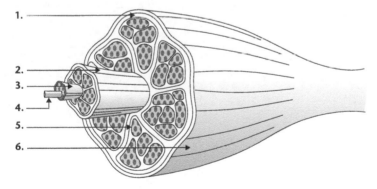

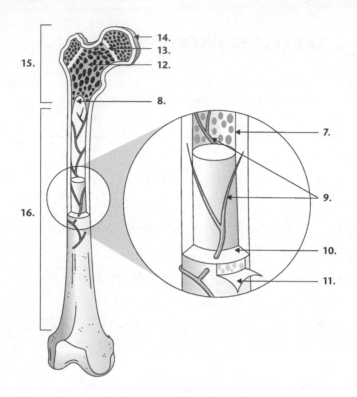

TRUE OR FALSE?

Are the following statements true or false?

17 Muscle is a type of connective tissue.

18 There are five types of bone.

19 Haversian canals are a key feature of compact bone.

20 Sensory neurones initiate the stimulus that allows a muscle to contract.

21 Muscle contraction occurs via the sliding filament mechanism.

22 The inability of a muscle to maintain its contraction strength is called muscle fatigue.

23 Cardiac and smooth muscles have the same properties.

24 Skeletal muscle is under voluntary control.

25 Osteocytes are involved in bone resorption.

26 Ossification can be broken down into three component parts.

27 There are three classes of joints based on anatomical characteristics.

28 A hinge joint is an example of a synovial joint.

29 The patella is a type of flat bone.

30 The pelvic girdle consists of two hip bones, which are fused anteriorly by the pubic symphysis.

MATCH THE TERMS

Match each joint with its correct functional description.

A. Slight movement (amphiarthrosis)

B. Free movement (diarthrosis)

C. Almost immovable (synarthrosis)

31 Skull sutures _____

32 Shoulder _____

33 Elbow _____

34 Pubic symphysis _____

35 Knee _____

36 Interphalangeal joints _____

37 Hip _____

38 Vertebrae _____

ANSWERS

 LABELLING EXERCISE

Figure 4.2 Structures and regions of bone and muscle

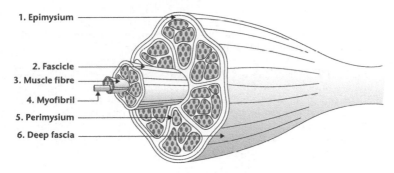

1. Epimysium
2. Fascicle
3. Muscle fibre
4. Myofibril
5. Perimysium
6. Deep fascia

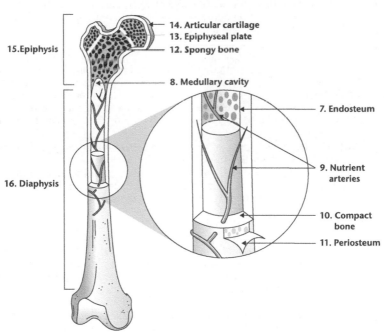

15. Epiphysis

16. Diaphysis

14. Articular cartilage
13. Epiphyseal plate
12. Spongy bone

8. Medullary cavity

7. Endosteum

9. Nutrient arteries

10. Compact bone

11. Periosteum

1 **Epimysium:** a layer of connective tissue surrounding a muscle. It is composed of dense irregular connective tissue and is continuous with fascia, endomysium and perimysium. It is also continuous with tendons where it becomes thicker and collagenous. The epimysium also protects muscles from friction between other muscles and bones.

2 **Fascicle:** a bundle of muscle fibres surrounded by a protective layer of perimysium.

3 **Muscle fibre:** skeletal muscle is a form of striated muscle tissue under control of the somatic nervous system. It is one of three major muscle types, the others being cardiac and smooth muscle. Skeletal muscle is made up of individual muscle fibres. The term muscle refers to bundles of muscle fibres held together by connective tissue.

4 **Myofibril:** cylindrical organelles found within muscle cells. They are bundles of actomyosin filaments (actin and myosin) that run from one end of the cell to the other and are attached to the cell surface membrane at each end. Actomyosin filaments are important in muscle contraction during which they slide past each other until completely overlapped but do not change length. This is known as the 'sliding filament theory' of muscle contraction.

5 **Perimysium:** a layer of dense connective tissue that surrounds muscle fibres (fascicles). The endomysium penetrates into each fascicle and separates individual muscle fibres.

6 **Deep fascia:** muscle fascia is a dense connective tissue that holds muscles together separating them into functional groups. The deep fascia of muscle allows free movement of muscles, supports nerves, blood and lymphatic vessels and fills intramuscular space. Three layers of dense irregular connective tissue extend from the deep fascia to protect and strengthen skeletal muscle: the epimysium, perimysium and endomysium.

7 **Endosteum:** a thin layer of connective tissue which lines the surface of bony tissue within the medullary cavity of long bones. It is usually resorbed during long periods of malnutrition. The outer surface of bone is covered with a thin layer of connective tissue (periosteum) which is structurally and functionally similar to endosteum.

8 **Medullary cavity:** the central cavity of the bone shaft where red bone marrow is stored. Located in the main diaphysis of long bone, which consists mostly of compact bone, the medullary cavity has walls composed of spongy bone and is lined with endosteum. The clavicle (collarbone) is the only long bone that does not contain a medullary cavity. The medullary cavity is mainly involved in the formation of erythrocytes and leucocytes.

9 **Nutrient artery:** a large artery that supplies the medullary canal in long bone. These arteries pass through a large opening, the nutrient foramen, which runs obliquely through the shaft to the medullary canal.

Long bone is well supplied with arterial networks. Many smaller arteries pierce compact bone to supply the spongy bone and red marrow while numerous arterioles run into the compact bone to supply the Haversian canals and systems.

10 *Compact bone:* a densely packed bone that contains few spaces. It forms the external layer on all bones and the bulk of diaphyses of long bones. Compact bone tissue provides protection and support and helps long bones to resist stress.

Blood vessels, lymphatics and nerves from the periosteum penetrate the compact bone through a system of perforating canals. The blood vessels and nerves of these canals connect with those of the medullary cavity, periosteum and Haversian canals.

11 *Periosteum:* a membrane that lines the outer surface of all bones, except at the joints of long bones. It is similar in structure to endosteum which lines the inner surface of all bones. Periosteum is attached to bone by strong collagenous fibres; it provides attachment for muscles and tendons and nourishment via the blood supply. Periosteum has sensory nerve endings which cause pain when damaged. During fracture healing, osteoblasts form bone and chondroblasts form cartilage, both are essential for healing.

12 *Spongy bone:* (cancellous bone) located in the epiphyses of long bone and the centre of all other bone types. It consists of a lattice structure of bone rods called trabeculae. The spaces between the trabeculae are filled with red marrow.

13 *Epiphyseal plate:* a hyaline cartilage plate in the metaphysis at each end of a long bone. This plate is found in children and adolescents, while in adults, it is replaced by an epiphyseal line because adults are no longer growing. At the end of puberty, the epiphyseal cartilage cells stop duplicating and the entire cartilage is slowly replaced by bone, leaving only a thin epiphyseal line.

14 *Articular cartilage:* a type of hyaline cartilage that is attached to articular bone surfaces. Articular cartilage does not bind bones together but reduces friction at the joint when bones move. It also absorbs shock.

15 *Epiphysis:* the round end of a long bone at its joint with adjacent bones. Between the epiphysis and diaphysis (the long midsection of the long bone) lies the metaphysis, including the epiphyseal (growth) plate.

16 *Diaphysis:* the main or mid-section of long bone. It is made up of compact bone and usually contains bone marrow and adipose tissue.

TRUE OR FALSE?

17 **Muscle is a type of connective tissue.**

Muscle is a type of tissue distinct from epithelial, connective and neural tissues. There are three kinds of muscle tissue (skeletal, cardiac and smooth) which differ from each other in terms of their microscopic anatomy, location and control by the nervous and endocrine systems.

Skeletal muscle is voluntary muscle tissue attached primarily to bones which facilitates movement of the skeleton. Cardiac muscle forms part of the heart, it is an involuntary muscle and is not part of the musculoskeletal system. Smooth muscle tissue is located in the walls of hollow internal structures including blood vessels and the intestinal tract. It is usually involuntary muscle with some automatic rhythm; it may also be influenced by hormones and neurotransmitters. Again, it is not part of the musculoskeletal system.

18 **There are five types of bone.**

Most bones of the body may be classified into five types:

I. Long bones are longer than they are wide. They are slightly curved which gives them added strength. Long bones consist primarily of compact bone tissue which is dense with few spaces. They also contain considerable amounts of spongy bone which has numerous large spaces. Long bones include the tibia, fibula, humerus, radius and ulna. The phalanges of the fingers and toes are also considered long bones.

II. Short bones are equal in length and width. They have a thin outer surface of compact bone. The carpals (in hands) and tarsals (in feet) are examples of short bones.

III. Flat bones are thin and composed of two plates of compact bone enclosing a layer of spongy bone. Flat bones protect the internal organs and provide areas for muscle attachment. Flat bones include the sternum, ribs and scapulae.

IV. Irregular bones have complex shapes. The amount of compact and spongy bone varies. Irregular bones include the vertebrae and some facial bones.

V. Sesamoid bones are generally embedded in tendons where considerable pressure develops. The patella (kneecap) is a sesamoid bone.

19 **Haversian canals are a key feature of compact bone.**

Haversian canals (osteons) are a series of tubes around narrow channels formed by lamellae in compact bone. Osteons are arranged parallel to the long axis of the bone. This arrangement allows the deposit of mineral salts and storage which give bone tissue its strength.

20 **Sensory neurones initiate the stimulus that allows a muscle to contract.**

Sensory neurones conduct nerve impulses to the central nervous system. Motor neurones stimulate muscle contraction. The motor neurone and muscle fibre are connected at the neuromuscular junction where muscle fibre membranes form specialized motor end plates. When an impulse from the brain reaches the axonal end of a motor neurone, some of the vesicles release neurotransmitters into the gap that lies between the motor end plate and the neurone. This causes the muscle fibre to contract. A single motor neurone may synapse with one or more muscle fibres.

21 **Muscle contraction occurs via the sliding filament mechanism.**

All muscle cells are made of actin and myosin filaments. The basic unit of organization of these contractile proteins in striated muscle cells is called the sarcomere. When each end of the myosin filament moves along the actin filament with which it overlaps, the two actin filaments are drawn closer together. This is the sliding filament theory of muscle contraction (see Answer 4). Thus, the ends of the sarcomere are drawn in and the sarcomere shortens although the length of actin and myosin filaments does not change. Sarcomeres are arranged in series; when a muscle fibre contracts, all sarcomeres within the fibre contract simultaneously.

22 **The inability of a muscle to maintain its contraction strength is called muscle fatigue.**

When skeletal muscles become over-stimulated, the strength of contraction becomes progressively weaker until the muscle no longer responds. The inability of a muscle to maintain its strength of contraction or tension is called muscle fatigue. It occurs when a muscle cannot produce enough energy to meet its needs. Several factors contribute to muscle fatigue, including insufficient oxygen, low glycogen levels, accumulation of lactic acid or a failure in neurotransmitter release.

23 **Cardiac and smooth muscles have the same properties.**

Cardiac muscles differ from smooth muscle as they are smaller and are 'branched' (that is, has the ability to connect to other cells). Each cardiac muscle cell (myocyte) connects with other cells at structures called intercalated discs. The gap junction between cells ensures that the movement of ions and molecules facilitates rapid propagation of nerve impulses. Cardiac muscle also has the unique skill of autorhythmicity (the ability to generate a contraction without input from the central nervous system). Conductivity (the ability to rapidly transfer the action potential from cell to cell) can occur as rapidly as one metre per second and contractility (the ability of the myocytes to convert electrical stimulus into mechanical contraction) occurs without cells reverting to an anaerobic state, thus allowing continual stimulation of these cells.

24 **Skeletal muscle is under voluntary control.**

Voluntary contraction of the skeletal muscles is used to move the body; it can be finely controlled. Examples include eye movements and movement of quadricep muscle of the thigh. There are two broad types of voluntary muscle fibres: slow twitch and fast twitch. Slow twitch fibres contract for long periods of time but with little force while fast twitch fibres contract quickly and powerfully but fatigue very rapidly.

Muscles are predominately powered by the breakdown of fats and carbohydrates, but anaerobic chemical reactions are also used, particularly by fast twitch fibres. These chemical reactions produce ATP which is used to power the muscle movement.

25 **Osteocytes are involved in bone resorption.**

There are four unique types of cell in bone tissue: osteoprogenitor cells, osteoblasts, osteocytes, and osteoclasts. Each has a specific function within the bone matrix.

Osteoprogenitor cells are non-specialized cells found in the inner part of the periosteum, in the endosteum and in the canals of bone that possess blood vessels. They undergo mitosis and develop into osteoblasts.

Osteoblasts are the cells that are involved in bone formation but are incapable of cell division (mitosis). These cells secrete collagen and other organic substances required in the formation of bone tissue.

Osteocytes are mature bone cells derived from osteoblasts. The osteocytes are the principal cells of bone tissue and, as with osteoblasts, cannot divide by mitosis. Osteoclasts are found on the surfaces of bone and function in bone resorption.

26 **Ossification can be broken down into three component parts.**

Ossification (osteoneogenesis) can be broken down into two component parts: intramembranous ossification and endochondral ossification. Intramembranous ossification occurs in foetal and embryonic fibrous connective tissue, where the mesenchymal cells come together (known as the centre of ossification), harden (through the deposit of calcium and other mineral salts), form the trabeculae (with blood vessels interwoven through it), and then develop the periosteum. Endochondral ossification is where existing hyaline cartilage develops first, which is then replaced by bone through development of a primary and secondary ossification centre, and then finally though formation of articular cartilage and an epiphyseal plate.

27 **There are three classes of joints based on anatomical characteristics.**

The structural classification of joints is based on the presence or absence of a space (synovial cavity) between the articulating (neighbouring) bones and the type of connective tissue that binds bones together.

A fibrous joint has no synovial cavity and the bones are held together by fibrous connective tissue. In cartilagenous joints, there is no synovial cavity but the bones are held together by cartilage. Synovial joints possess a joint cavity and the bones forming the joint are united by a surrounding articular capsule and in some cases by accessory ligaments.

Joints may also be classified by function and the degree of movement they permit. A synarthrosis joint is immovable, as with most fibrous joints. Amphiarthroses are slightly movable so most cartilagenous joints are in this category. Diarthroses are freely movable joints (such as shoulder or hip) so all synovial joints fall under this description.

28 **A hinge joint is an example of a synovial joint.**

A synovial joint is the most common and most movable joint in the body. As with most other joints, synovial joints achieve movement at the point of contact with the articulating bones. Structural and functional differences distinguish synovial joints from cartilaginous and fibrous joints. The main structural differences are: (1) capsules surrounding the articulating surfaces of a synovial joint; and (2) the lubricating synovial fluid within the capsule cavity. Examples of synovial joints include the elbow and ball-and-socket hip joint.

29 **The patella is a type of flat bone.**

The patella (kneecap) is a sesamoid bone – the largest in the body. It is a thick, circular-triangular bone which articulates with the femur and covers and protects the knee joint. The primary function of the patella is knee extension.

30 **The pelvic girdle consists of two hip bones, which are fused anteriorly by the pubic symphysis.**

The two hip bones (otherwise called the coxal bones) are fused by the pubic symphysis anteriorly and join with the sacroiliac joint posteriorly. The two hip bones each consist of three separate parts in the newborn (superior ilium, inferior and anterior pubis, and an inferior and posterior ischium) which fuse to become one bone (although are still described in adulthood as three different areas). Furthermore, the bony pelvis can be divided into superior and inferior portions by a boundary known as the pelvic brim. Above the pelvic brim is the areas known as the false, or greater pelvis, where the area inferior to the pelvic brim is known as the true, or lesser pelvis.

MATCH THE TERMS

31	Skull sutures	**C.** Almost immovable (synarthrosis)
32	Shoulder	**B.** Free movement (diarthrosis)
33	Elbow	**B.** Free movement (diarthrosis)
34	Pubic symphysis	**A.** Slight movement (amphiarthrosis)
35	Knee	**B.** Free movement (diarthrosis)
36	Interphalangeal joints	**B.** Free movement (diarthrosis)
37	Hip	**B.** Free movement (diarthrosis)
38	Vertebrae	**A.** Slight movement (amphiarthrosis)

5 The nervous system and special senses

INTRODUCTION

Cellular communication is controlled by two separate but interconnected systems – the nervous and endocrine systems. The nervous system produces rapid, short-term responses; endocrine responses are slower and generally longer-term. The nervous system gathers, transfers and processes information from the brain, spinal cord and peripheral nerves. It can be divided into the central and peripheral nervous systems. The central nervous system (CNS) consists of the brain and spinal cord while the peripheral nervous system (PNS) links all limbs and organs with the CNS. The PNS can be further divided into the autonomic and somatic nervous systems. The autonomic system is additionally subdivided into the sympathetic and parasympathetic systems.

Neurones are the main cells of the nervous system and function in coordination and communication throughout the body. Neuroglia (glial cells) are nervous system support cells which regulate the environment around neurones.

The nervous system helps regulate homeostasis and integrates all body functions by sensing changes (sensory), interpreting them (integration) and reacting to them (motor activities).

Paramedics should know that the organization of the nervous system is fundamental to understanding how different parts of the body communicate and how the body responds to stimuli to maintain homeostasis and prevent illness.

Useful resources

Paramedics! Test Yourself in Pathophysiology
Chapter 5

Anatomy and Physiology (8th edition)
Chapters 11, 12 and 13

BBC: The human body systems:
http://www.bbc.co.uk/science/humanbody/body/index.shtml?nervous

Brain Atlas: How the brain works:
http://www.brainexplorer.org/neurological_control/Neurological_index.shtml

---

LABELLING EXERCISE

1–15 Identify the features of a neurone and spinal reflex arc in Figure 5.1, using the terms provided in the box below.

axonal terminal	white matter
node of Ranvier	grey matter
myelin sheath	relay neurone
axon	central canal
cell body	synapse
dendrites	motor neurone
receptor organ	effector organ
sensory neurone	

Figure 5.1 The neurone and spinal reflex arc

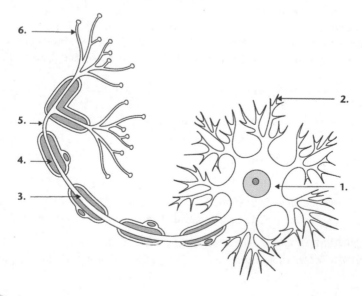

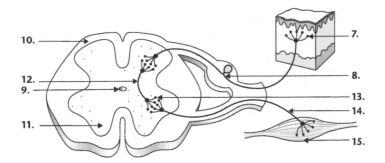

TRUE OR FALSE?

Are the following statements true or false?

16 The central nervous system (CNS) consists of the brain and spinal cord.

17 There are four types of neurone.

18 Neuroglia are the supporting cells in the CNS.

19 The peripheral nervous system is subdivided into the autonomic and the somatic nervous systems.

20 The autonomic nervous system (ANS) controls voluntary activities.

21 The cell body of the neurone acts as a receptor site for nerve impulses.

22 Saltatory conduction is much faster than nerve impulses that move along unmyelinated membranes.

23 The nervous system deteriorates with age.

24 Important bodily functions, such as respiration and digestion, are controlled through reflexes.

25   Taste and smell are electrically stimulated senses.

26   There are two layers of meninges.

27   There are three basic skin sensations.

28   Cerebrospinal fluid is produced by the arachnoid villi.

MULTIPLE CHOICE

Identify one correct answer for each of the following.

29 A typical neural pathway is:

  a) effector-sensory nerve-PNS-motor nerve-receptor
  b) receptor-sensory nerve-CNS-motor nerve-effector
  c) effector-motor nerve-PNS-sensory nerve-receptor
  d) receptor-motor nerve-CNS-sensory nerve-effector

30 In most reflex arcs, sensory neurones synapse in:

  a) the dura mater
  b) the cerebrospinal fluid
  c) the spinal cord
  d) the brain stem

31 The action potential is described as being:

  a) one-for-all
  b) all-for-one
  c) nothing-at-all
  d) all-or-nothing

32 Each cerebral hemisphere is divided into how many lobes?

  a) 2
  b) 3
  c) 4
  d) 5

33 How many pairs of cranial nerves originate in the brain?

a) 8

b) 10

c) 12

d) 14

34 The spinal cord is continuous with which region of the brain?

a) cerebrum

b) medulla oblongata

c) midbrain

d) pons

35 The sympathetic nervous system mediates:

a) rest and digest responses

b) fright, fight or flight responses

c) sleeping responses

d) gastric responses

36 Nerve impulses from visual stimuli are integrated in which lobe?

a) frontal

b) temporal

c) parietal

d) occipital

37 In the eye, light rays are refracted onto:

a) the cornea

b) the lens

c) the retina

d) the blind spot

38 Which of the following is not an anatomical structure of the ear?

a) tympanic membrane
b) olfactory organs
c) semi-circular canals
d) cochlea

39 The two enlargements of the spinal cord are known as:

a) the cervical and thoracic enlargements
b) the cervical and lumbar enlargements
c) the thoracic and lumbar enlargements
d) the lumbar and sacral enlargements

FILL IN THE BLANKS

Fill in the blanks in each statement using the options in the box below.
Not all of them are required, so choose carefully!

anaesthetics	noradrenaline
light adaptation	dark adaptation
cerebellum	cerebrum
reflex	synapse
proprioceptors	adrenaline
taste buds	inhibitory
nostrils	blood–brain barrier
dermatomes	

40 Neurones are connected by a _____.

41 A _____ arc is the neural pathway that mediates nervous activity.

42 The largest part of the brain is the _____.

43 The _____ is situated below the occipital lobes of the cerebrum.

44 _____ neurones in the CNS block unimportant signals and permit transmission of selected information.

45 _____ block nerve impulses by reducing membrane permeability to sodium ions.

46 _____ is the neurotransmitter that regulates the sympathetic nervous system.

47 _____ are sense organs stimulated by body movement.

48 When a person is dazzled after passing from darkness into light, the adjustment period is called _____ _____.

49 The gustatory receptors are located in the _____ _____.

50 The _____ _____ _____ protects the brain from potentially harmful elements in the blood.

51 _____ are areas of the skin that provide sensory information to the CNS.

ANSWERS

 LABELLING EXERCISE

Figure 5.2 The neurone and spinal reflex arc

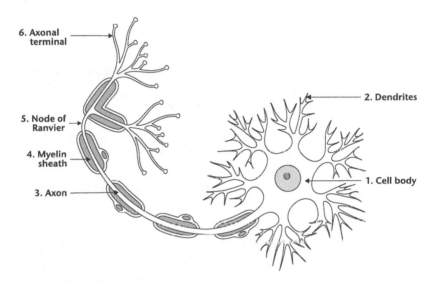

6. Axonal terminal
2. Dendrites
5. Node of Ranvier
4. Myelin sheath
3. Axon
1. Cell body

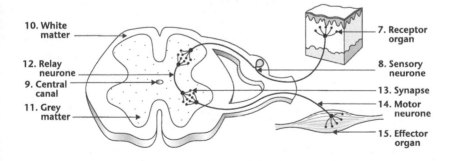

10. White matter
7. Receptor organ
12. Relay neurone
8. Sensory neurone
9. Central canal
13. Synapse
11. Grey matter
14. Motor neurone
15. Effector organ

1   **Cell body (or soma):** every nerve cell (neurone) has a cell body which contains a nucleus and granular cytoplasm containing many ribosomes. These ribosomes are grouped together to form Nissl granules which are involved in the formation of the neurotransmitter substances. It is the high concentration of Nissl granules in the cell bodies that gives parts of the brain and spinal cord a grey colour (hence the term 'grey matter').

2   **Dendrites:** these are the fine branches which provide a large connection network, capable of receiving nerve impulses (in the form of an electrical signal called an action potential) from neighbouring nerve cells and receptor organs. The dendrites make connections with other nerve cells or effector organs (such as muscles) via synapses. Neurotransmitter substances pass across the synapses to neighbouring cells and hence the nerve impulse is transmitted from one neurone to the next.

3   **Axon:** the axon is the communication route between the cell body and the axon terminals. The action potential will only travel in one direction along the axon. In healthy neurones, the axons are wrapped in a segmented fatty deposit called the myelin sheath.

4   **Myelin sheath:** the myelin sheath is produced by neuroglia (see Answer 18) called Schwann cells. Myelin is a fatty mixture of phospholipid and cholesterol which provides insulation from electrical impulses. Development of the myelin sheath is not complete until late childhood hence the slow responses and poor coordination often exhibited by infants and children.

5   **Nodes of Ranvier:** these are small unmyelinated regions located along the axis between the myelin segments. Due to the absence of myelin, these regions are not insulated and therefore can conduct the electrical impulse from one node of Ranvier to the next. This is why the electrical impulse is said to jump along the axon – it cannot travel smoothly along the myelinated axon and so it is only conducted at the nodes of Ranvier. The jumping of the action potential from one node to the next is called saltatory conduction.

6   **Axonal terminals:** the end of the axon consists of numerous projections which are involved in the transmission of the action potential to the dendrites of neighbouring neurones or the effector organ. Each has a synaptic end plate which stores neurotransmitter substance for release into the synapse.

7   **Receptor organ:** the receptor organ receives a stimulus from the environment, therefore is the origin of any reflex action.

8   **Sensory (afferent) neurone:** conveys the nerve impulse, in the form of an action potential, from the receptor to the reflex centre in the CNS. Sensory neurones are myelinated, therefore transmission is rapid.

9 **Central canal:** the centre of the spinal cord. It contains cerebrospinal fluid (CSF) to nourish and protect the CNS. CSF performs the same function in the subarachnoid space of the meninges in the brain.

10 **White matter:** contains many motor and sensory neurones which are well myelinated. The fatty myelin is white in colour compared to grey matter. In the spinal cord, white matter is on the exterior of the cord while grey matter is in the inner region.

11 **Grey matter:** the absence of myelin on the neurones and glial cells in this region is responsible for the grey colour. In the brain, grey matter forms the outer layer of the cortex while in the spinal cord grey matter is located on the inner area.

12 **Relay neurone (interneuron):** conveys nerve impulses from sensory neurones across the spinal cord. They are unmyelinated and therefore impulse transmission is relatively slow. The whole neurone (cell body to axon terminal) is contained within the CNS.

13 **Synapse:** all neurones are linked to neighbouring neurones by synapses which also connect neurones to receptor and effector organs. Synapses are located between the axon terminals of one neurone and the dendrites of neighbouring neurones. One neurone can synapse with many neighbouring neurones which is why nerve impulses travel very quickly. In the reflex arc, sensory neurones synapse with relay neurones, which synapse with motor neurones.

14 **Motor (effector) neurones:** the nerve cells that transmit the body's response to the stimulus.

15 **Effector organ:** responds to stimuli by performing an action after receiving a nerve impulse via a motor neurone. Examples are muscles or glands.

TRUE OR FALSE?

16 **The central nervous system (CNS) consists of the brain and spinal cord.**

The principal parts of the brain are the brain stem, cerebellum, diencephalon and cerebrum. The spinal cord consists of a mass of nervous tissue located in the vertebral canal from which 31 pairs of spinal nerves originate.

The CNS is protected from physical shock and infection by the bones of the skull and the vertebral column. It is also protected by the cushioning effects of the cerebrospinal fluid (CSF) and the meninges (the dura mater, arachnoid and pia mater). Unlike the CNS, the peripheral nervous system (PNS) is not protected by bone or the blood–brain barrier. It consists of all neural tissue outside the CNS. Its main function is to connect the limbs and organs to the CNS. It can be subdivided into the afferent division and efferent division. The afferent division carries sensory information to the CNS while the efferent division transmits motor information away from the CNS to the muscles and glands.

17 **There are four types of neurone.**

There are three types of neurone:

I. Sensory neurones – carry nerve impulses from the receptor cells of sensory organs to the CNS. These neurones usually have myelinated axons that facilitate rapid transmission of the nerve impulse.

II. Motor neurones – conduct impulses away from the CNS to the effectors which include the muscles and glands. They are not myelinated in the autonomic nervous system but are myelinated in the somatic nervous system.

III. Relay neurones – a connecting neurone between sensory and motor neurones. They are myelinated in the white matter of the CNS but non-myelinated in the grey matter of the CNS. Most (but not all) reflex arcs include relay neurones. For example, the knee-jerk reflex involves no relay neurones.

18 **Neuroglia are the supporting cells in the CNS.**

Approximately 40 per cent of the brain consists of four types of neuroglia that support neurones by supplying nutrients, producing myelin and CSF and removing toxic waste.

19 **The peripheral nervous system is subdivided into the autonomic and the somatic nervous systems.**

The PNS forms a network throughout the body supplying motor nerves to individual effector organ cells and receiving nerves from individual sensory cells. The sensory division of the PNS conducts information from sensory organs (receptors) to the CNS. The motor division of the PNS conducts nerve impulses from the CNS to the muscles and glands (effectors). The motor division can be further divided into the autonomic (involuntary) and the somatic (skeletal or voluntary) nervous systems. These systems generally work antagonistically, where the action of one counteracts the action of the other (see Figure 5.3).

Figure 5.3 Structural and physiological divisions and interactions in the nervous system

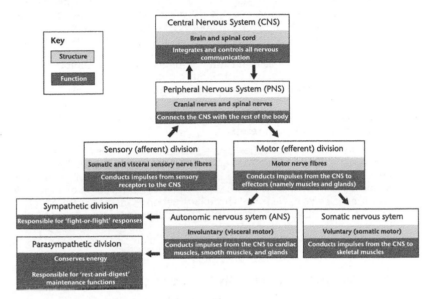

20　The autonomic nervous system (ANS) controls voluntary activities.

The autonomic (or visceral motor) nervous system controls involuntary functions mediated by the activity of smooth muscle fibres, cardiac muscle fibres and glands.

21　The cell body of the neurone acts as a receptor site for nerve impulses.

The dendrites receive nerve impulses which are conducted to the cell body and transmitted along the axon to the axonal dendrites from which the nerve impulse is then transmitted to neighbouring neurones.

22　Saltatory conduction is much faster than nerve impulses that move along unmyelinated membranes.

Saltatory conduction describes the 'jumping/hopping' movement of action potentials along myelinated axons from one node of Ranvier to the next. This 'jumping' mechanism increases the speed of action potentials in myelinated neurones compared to unmyelinated neurones. Depolarization at one node of Ranvier is sufficient to trigger an action potential at the next node. Thus in myelinated axons, action potentials do not move as waves, but occur at successive nodes, effectively 'jumping' along the axon.

23 **The nervous system deteriorates with age.**

Anatomical and physiological changes begin in the nervous system shortly after maturity (around age 30 years) and accumulate over time. Common age-related anatomical changes include decrease in brain size and weight coupled with a reduction in the number of neurones and synapses. A number of intracellular changes may occur in the neural cells and blood supply to the brain may be diminished. Any combination of these changes is associated with impaired neural function and is linked with neurodegenerative conditions such as Alzheimer's disease.

24 **Important bodily functions, such as respiration and digestion, are controlled through reflexes.**

There are two types of reflex arc – the autonomic reflex arc (affecting inner organs) and the somatic reflex arc (affecting muscles). Some reflex arcs occur without conscious knowledge such as change in pupil size or gastric movements, while there is conscious awareness of others. A conditioned reflex is 'learned' through experience such as the production of saliva upon the sight or smell of food. An unconditioned (inborn) reflex happens directly upon nervous stimulation such as the production of saliva when food enters the mouth and stimulates sensory receptors in the tongue.

25 **Taste and smell are electrically stimulated senses.**

Taste and smell are chemically stimulated senses. To stimulate smell, a substance must be in a gaseous form and then converted to a solution. To stimulate taste, a substance must be in solution to trigger the taste buds.

26 **There are two layers of meninges.**

There are three main layers of the meninges: the dura mater (outer layer), the arachnoid mater (middle layer) and the pia mater (the inner layer). Their function is to cover and protect the brain, allowing CSF to occupy the inner layers of the meninges and act as a 'shock absorber'.

27 **There are three basic skin sensations.**

There are five basic skin sensations: touch, pressure, pain, warmth, cold. On hairy skin surfaces, nerve fibres around the hair shaft register touch sensation when hair moves. On hairless skin surfaces (for example, palms of hands, soles of feet), various types of nerve endings exist to detect touch, pressure and cold. Pain is detected by nerve endings in the epidermis and dermis. Other sensations such as tickling, itching, softness, hardness and wetness are thought to be due to stimulation of two or more of these special nerve endings and to blending of these sensations in the brain.

28 **Cerebrospinal fluid is produced by the arachnoid villi.**

Cerebrospinal fluid (CSF) is reabsorbed into the blood by the arachnoid villi, but is produced by the choroid plexuses of each ventricle. An adult has between 80–150mls of CSF, which offers mechanical protection (shock absorber) and chemical protection (helping neurone signalling). The CSF also acts as an exchange route for nutrients and waste products.

 MULTIPLE CHOICE

Correct answers identified in bold italics.

29 **A typical neural pathway is:**

a) effector-sensory nerve-PNS-motor nerve-receptor   *b) receptor-sensory nerve-CNS-motor nerve-effector*   c) effector-motor nerve-PNS-sensory nerve-receptor   d) receptor-motor nerve-CNS-sensory nerve-effector

This is the pathway from stimulus of the receptor such as an accidental pin prick to the finger. Receptor cells detect the injury and transmit a nerve impulse along sensory neurones to the CNS which integrates the impulse (sometimes via a relay neurone). The CNS then sends a response via motor neurones to the effector organs (such as the muscles) to initiate a response such as remove finger from the pin.

30 **In most reflex arcs, sensory neurones synapse in:**

a) the dura mater   b) the cerebrospinal fluid   *c) the spinal cord*   d) the brain stem

This allows reflex actions to occur relatively quickly by activating spinal motor neurones without the delay of sending signals to the brain – although the brain will receive sensory input while the reflex action occurs.

31 **The action potential is described as being:**

a) one-for-all   b) all-for-one   c) nothing-at-all   *d) all-or-nothing*

When the nerve fibre is resting, it is permeable to potassium ions (K^+) located inside the cell and relatively impermeable to the sodium ions (Na^+) outside the cell. Inside the nerve fibre is negatively charged relative to the outside because the positive charge of the Na^+ is more influential than the K^+. In this resting state the nerve fibre is said to be polarized (that is, a difference in electrical charge exists). A nerve fibre becomes stimulated when an electrical current reduces the difference between the electrical charges inside and outside the neurone, this is called depolarization. This depolarization alters the membrane permeability, allowing Na^+ to enter the fibre, making inside the fibre positively charged. The nerve impulse is all-or-nothing meaning that stimuli must be large enough to produce sufficient depolarization to trigger the event. The size of the action potential is dependent on the concentration of Na^+ and K^+, not the size of the stimulus. Any stimulus that does not produce enough depolarization to allow Na^+ to enter the cell will not be transmitted – hence, all-or-nothing (see Figure 5.4).

Figure 5.4 The action potential

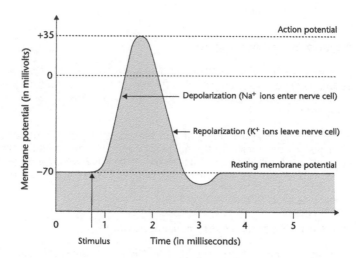

32 Each cerebral hemisphere is divided into how many lobes?

a) 2   b) 3   *c)* 4   d) 5

The cerebrum is the largest part of the brain. It is divided into two hemispheres which are each further sub-divided into four lobes. These lobes – frontal, temporal, parietal and occipital lobes – are named after the cranial bones that lie over them. Each lobe has anatomical and functional differences. The frontal lobe influences personality, judgement, abstract reasoning, social behaviour, language expression and voluntary movement. The temporal lobe controls hearing, understanding of language, learning, storage and recall of memories. The parietal lobe interprets and integrates sensation, inducing pain, temperature and touch. It also interprets size, shape, distance, vibration and texture. The occipital lobe mainly interprets visual stimuli from the eyes.

33 | **How many pairs of cranial nerves originate in the brain?**

a) 8   b) 10   *c) 12*   d) 14

The 12 cranial nerves carry information to and from the brain and supply the head, neck and most of the viscera (that is, internal organs, specifically those within the chest or abdomen). The 12 nerve pairs (from front to back) are: olfactory (sensory nerve of smell); optic nerve; oculomotor; trochlear; trigeminal; abducens; facial; acoustic (auditory); glossopharyngeal; vagus; (spinal) accessory; hypoglossal.

The PNS also consists of 31 pairs of spinal nerves that carry information between the spinal cord and the trunk and limbs. These spinal nerves can be grouped as: 8 cervical pairs (C1 to C8), 12 thoracic pairs (T1 to T12), 5 lumbar (L1 to L5), 5 sacral (S1 to S5) and 1 coccygeal. The sacral and the coccygeal together form the cauda equina.

34 | **The spinal cord is continuous with which region of the brain?**

a) cerebrum   *b) medulla oblongata*   c) midbrain   d) pons

The spinal cord joins the medulla oblongata at an opening in the occipital portion of the skull. The medulla oblongata (or medulla) is part of the brain stem – along with the midbrain and pons. It influences cardiac, respiratory and vasomotor functions by housing the medullary rhythmicity area, as well as the cardiac centre. It is also the centre for vomiting, coughing and hiccupping reflexes. Some of the white matter of the medulla tracts form protrusions on the anterior aspect which are known as the pyramids. The cross-over of motor tracts from the left to right and right to left is known as the decussation of pyramids.

35 | **The sympathetic nervous system mediates:**

a) rest and digest responses   *b) fright, fight or flight responses*
c) sleeping responses   d) gastric responses

The sympathetic nervous system is part of the autonomic nervous system that increases the body's ability to be active or react to emergencies, hence the phrase 'fright, fight or flight'. It increases heart rate, dilates bronchi, constricts blood vessels therefore increasing blood pressure and reduces gastric activity by constricting sphincter muscles. The parasympathetic nervous system is called the 'rest and digest' system as it produces the opposite effects of the sympathetic system. It is associated with maintaining normal homeostasis and is therefore said to be responsible for 'housekeeping'.

36 | **Nerve impulses from visual stimuli are integrated in which lobe?**

a) frontal   b) temporal   c) parietal   *d) occipital*

The occipital lobe houses the visual processing centre and interprets visual stimuli. If stimulus is received via the left retina, it will be interpreted by the right occipital lobe because one side of the occipital cortex receives impressions from the field of vision on the opposite side. The frontal lobe controls voluntary muscle movement. The temporal lobe contains the centre for taste, hearing and smell. The parietal lobe coordinates and interprets sensory information from the senses.

37 In the eye, light rays are refracted onto:

a) the cornea   b) the lens   *c) the retina*   d) the blind spot

The retina lies at the back of the eye and consists of the nervous tissue where the visual stimulus is received before transmission to the visual cortex in the brain. The retina has three layers: (1) photoreceptors (rods and cones) – to absorb light; (2) bipolar neurones – to generate action potentials; and (3) sensory neurones – to conduct nerve impulses to the brain via the optic nerve.

38 Which of the following is not an anatomical structure of the ear?

a) tympanic membrane   *b) olfactory organs*   c) semi-circular canals   d) cochlea

The olfactory organs are the organs of smell. They exist in the nasal cavity as a pair, on either side of the nasal septum. The ear has three different anatomical regions. These are: (1) the outer ear (auricle, pinna) which receives sound waves and sends these, as vibrations, to the tympanic membrane (ear drum); (2) the middle ear – contains three small bones (hammer, anvil and stirrup), the auditory ossicles. The Eustachian tube connects the middle ear to the nasopharynx; and (3) the inner ear – contains the organs of hearing and equilibrium. The cochlea has receptors for hearing while the semi-circular canals are responsible for balance and equilibrium.

39 The two enlargements of the spinal cord are known as:

a) the cervical and thoracic enlargements   *b) the cervical and lumbar enlargements*   c) the thoracic and lumbar enlargements
d) the lumbar and sacral enlargements

The spinal cord has two enlargements, in the cervical and lumbar areas. Inferior to that, the cord narrows at the conus medullaris, ending at the filum terminae. From there, the nerve fibres spread out over the sacrum; this is known as the cauda equina.

FILL IN THE BLANKS

40 **Neurones are connected by a _synapse_.**

This is the space between two neurones or between a neurone and effector organ. A neurotransmitter substance is released into the gap and diffuses across the space to stimulate the dendrites of the neighbouring neurone to propagate the nerve impulse by opening sodium channels to allow depolarization. If an insufficient quantity of neurotransmitter substance is released, the neighbouring nerve cell will not be stimulated and the nerve impulse will not be transmitted, hence the nerve impulse is described as being all-or-nothing. In most synapses of the PNS, the neurotransmitter substance is acetylcholine. In the sympathetic nervous system the neurotransmitter is often noradrenaline. A synapse permits transmission of the nerve impulse in only one direction.

41 **A _reflex_ arc is the neural pathway that mediates nervous activity.**

The reflex arc is the neural pathway that mediates a reflex response. A nervous reflex is an involuntary action caused by the stimulation of a sensory neurone or receptor. Reflexes form the basis of all CNS activity. Most sensory neurones do not pass directly into the brain, but synapse in the spinal cord. This characteristic allows reflex actions to occur relatively quickly by activating spinal motor neurones without the delay of sending signals via the brain, although the brain will receive sensory input while the reflex action occurs.

42 **The largest part of the brain is the _cerebrum (or forebrain)_.**

The cerebrum has two hemispheres which are each divided into four lobes. A deep fissure superficially separates the cerebral hemispheres but they are still connected deep in the cleft through the corpus callosum, which allows transmission of nerve signals between the two sides of the brain. The surface of the cerebrum has many folds (convolutions). The cerebrum is the centre for intellect, memory, language, consciousness, personality, sensory and motor control. It consists of the cerebral cortex, basal ganglia and corpus callosum. The cerebral cortex is the outer layer and houses motor areas that control voluntary movement, sensory areas that interpret sensory information and association areas that control learning and emotion, as well as connecting and integrating the motor and sensory regions. The basal ganglia modify and coordinate gross muscle movement and regulate muscle tone. The limbic system is involved in emotion and controls involuntary behaviour, essential for survival.

43 **The _cerebellum_ is situated below the occipital lobes of the cerebrum.**

The second largest region in the brain, the cerebellum lies behind the cerebrum, below the occipital lobes and has two hemispheres. It controls balance and equilibrium by coordinating muscle movement. The cerebellum is also involved in controlling more skilled, coordinated movements such as walking and maintaining posture.

44 *Inhibitory* neurones in the CNS block unimportant signals and permit transmission of selected information.

Without such mechanisms, centres of the brain would be over-stimulated by nerve impulses. Inhibitory neurones possess neurotransmitters that prevent excitation of their neighbouring neurone, suppressing their activity. In general, the activity of a neurone depends on the balance between the number of excitatory and inhibitory processes affecting it and these can occur simultaneously. Important inhibitory neurotransmitters found in the cerebral cortex include gamma aminobutyric acid (GABA) and the amino acid, glycine.

45 *Anaesthetics* block nerve impulses by reducing membrane permeability to sodium ions.

When an impulse passes along a nerve, Na^+ enters the nerve fibre and K^+ leaves it. When this polarization–depolarization action reaches the synapse between two neurones or between a neurone and effector organ, a chemical substance (neurotransmitter) is released. This crosses the synapse gap and forms a stimulus to initiate the next neurone (or effector). Anaesthetics prevent the transmission of nerve impulses by preventing Na^+ from entering the cell hence the impulse cannot be propagated and nerve signals (associated with pain, muscle movement, regulation of blood supply and other body functions) will be inhibited. This is why anaesthetics are used in surgery. Relatively high doses of anaesthetics will inhibit all forms of sensations and muscle contraction, while lower doses will selectively inhibit pain sensation but have little effect on muscle power.

46 *Noradrenaline (norepinephrine)* is the neurotransmitter that regulates the sympathetic nervous system.

Noradrenaline is synthesized from dopamine. As a neurotransmitter in the CNS and sympathetic nervous system, it is released from noradrenergic neurones when physiological changes are triggered by a stressful event. The actions of noradrenaline are induced via the binding to adrenergic receptors and its effects are alertness and arousal as described by 'fright–fight–flight'. Acetylcholine is the main neurotransmitter in the parasympathetic nervous system.

47 *Proprioceptors* are sense organs stimulated by body movement.

These are receptors found all over the body and make us aware of position and movement of the body in relation to its surrounding environment and body parts in relation to each other. General proprioceptors are found in skeletal muscles, tendons and joints.

48 When a person is dazzled after passing from darkness into light, the adjustment period is called *light adaptation*.

This describes the chemical changes that occur after the initial dazzling. As the dazzle wears off and normal vision begins to return, the eyes are adjusting to interpreting more colour in the bright environment, whereas in the darkness the eyes concentrate on distinguishing various degrees of light and dark. When moving from light into darkness, the eyes undergo dark adaptation.

49 **The gustatory receptors are located in the *taste buds*.**

Substances to be tasted must be in solution in saliva. Receptor potentials develop in the gustatory receptor cells, triggering the release of neurotransmitters, which gives rise to nerve impulses. Adaption to taste occurs quickly and threshold varies with the taste involved – sweet, sour, salt, bitter. Taste signals are processed in the parietal lobe of the cerebral cortex, the thalamus and the medulla oblongata.

50 **The *blood–brain barrier* protects the brain from potentially harmful elements in the blood.**

The blood–brain barrier is a tightly packed group of endothelial cells that inhibits the movement of certain harmful elements from the blood into the brain, yet still allowing oxygen, carbon dioxide, alcohol and glucose to pass freely into the brain. Other substances, such as creatinine, urea and proteins, also pass through the blood–brain barrier into the brain but at a slower rate.

51 ***Dermatomes* are areas of the skin that provide sensory information to the CNS.**

In addition to sensory information, dermatomes also allow impulses from the spinal cord to innervate skeletal muscles. Each of the nerves relays sensation (including pain) from a particular region of skin to the brain.

6 The endocrine system

INTRODUCTION

The tissues and organs of the endocrine system secrete hormones from glandular cells. Hormones act as chemical messengers and are synthesized from precursors in the blood. Once released by glands, hormones are transported in the blood to other parts of the body where they exert their effects on specific targets. Some glands secrete substances into interstitial fluid, but because they are rapidly degraded, they do not reach the blood. These substances are more accurately defined as local messengers and include autocrine secretions, which affect only the secreting cells and paracrine secretions, which only affect cells adjacent to the gland.

The functions of the endocrine system are often associated and integrated with the nervous system. Certain parts of the nervous system stimulate or inhibit the release of hormones, while hormones may promote or inhibit the generation of nerve impulses. Furthermore, some molecules act as hormones at some locations and as neurotransmitters at others. Since the endocrine system involves more consistent communication, it is also essential for maintaining homeostasis. It is important for paramedics to understand how the endocrine and nervous systems work together to maintain normal hormone and fluid levels in the body.

Useful resources

Paramedics! Test Yourself in Pathophysiology
Chapter 6

Anatomy and Physiology (8th edition)
Chapter 15

Hole's Essentials of Human Anatomy and Physiology (11th edition)
Chapter 11

Hormones, receptors and the endocrine system:

http://www.vivo.colostate.edu/hbooks/pathphys/endocrine/basics/index.html
http://www.emc.maricopa.edu/faculty/farabee/BIOBK/BioBookENDOCR.html

---

LABELLING EXERCISE

1-8   Identify the organs of the endocrine system in Figure 6.1, using the options provided in the box on p. 82.

Figure 6.1 The endocrine system

N.B. Figure not to scale

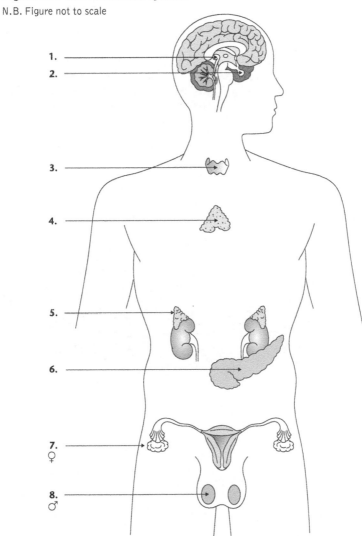

1.
2.
3.
4.
5.
6.
7.
♀
8.
♂

thymus gland

testis

ovary

adrenal gland

pineal gland

thyroid gland

pancreas

pituitary gland

TRUE OR FALSE?

Are the following statements true or false?

| **9** | The pancreas has both endocrine and exocrine functions. |

| **10** | Homeostasis of hormonal secretions is maintained only by negative feedback mechanisms. |

| **11** | The hypothalamus has a posterior and anterior lobe. |

| **12** | Oxytocin is produced by the pineal gland. |

| **13** | Body hair is produced in response to a follicle-stimulating hormone. |

| **14** | Oestrogens are present in males. |

| **15** | Secretion of antidiuretic hormone increases urine production. |

| **16** | The principal actions of thyroid hormones increase the basal metabolic rate. |

| **17** | Calcitonin is produced by the thyroid gland. |

| **18** | Bone resorption is facilitated by the action of the parathyroid hormone. |

19 The adrenal glands can be structurally and functionally differentiated into two distinct regions.

20 Glucocorticoids suppress the immune system.

21 The hormones produced by the adrenal medulla have major effects on heart rate and blood pressure.

22 All endocrine glands store the hormones they produce.

MULTIPLE CHOICE

Identify one correct answer for each of the following.

23 Melatonin is secreted by the:

a) pancreas

b) pineal gland

c) keratinocytes of the skin

d) ovaries

24 The thyroid gland:

a) is located inferior to the larynx

b) produces antidiuretic hormone

c) secretes small amounts of insulin

d) helps initiate milk production

25 Glucagon is produced in the pancreas by which cells?

a) alpha cells

b) beta cells

c) delta cells

d) F cells

26 Levels of which hormone are controlled by positive feedback?

a) growth hormone

b) thyroid-stimulating hormone

c) oxytocin

d) insulin

27 This hormone acts on the intestines and causes increased calcium absorption:

a) calcitonin

b) calcitriol

c) thyroxine

d) pancreatic polypeptide

28 Which hormone is secreted in response to high blood glucose?

a) insulin

b) glucagon

c) cortisol

d) oxytocin

29 Somatostatin is secreted by the:

a) pancreatic delta cells

b) pancreatic polypeptide cells

c) zona fasciculata

d) posterior pituitary

30 A lack of or decrease in insulin hormone receptors on cells can result in:

a) diabetes insipidus

b) Type 1 diabetes mellitus

c) Type 2 diabetes mellitus

d) juvenile diabetes

31 Which of the following characteristics is the same for the nervous and endocrine systems?

a) target cells affected

b) duration of actions

c) mechanism of signalling and communication

d) none of the above

32 The placenta produces:

a) oestrogen

b) progesterone

c) chorionic gonadotropins

d) all of the above

33 Which of the following is an example of a peptide hormone?

a) aldosterone

b) serotonin

c) gastrin

d) prostaglandins

MATCH THE TERMS

Identify which gland (A–H) is related to the following hormones and their responses (34–41).

A. Thyroid

B. Adrenals

C. Pituitary

D. Ovaries

E. Parathyroids

F. Hypothalamus

G. Pancreas

H. Testes

	Hormone	*Response*	*Gland*
34	Trophic hormones	Exert their effects on activities of pituitary gland. This gland also secretes two other hormones (ADH and oxytocin) that are stored in the pituitary for later release	___
35	Antidiuretic hormone	Enhances water conservation by the kidneys	___
36	Thyroxine	Regulates the metabolic rate and energy production	___
37	Parathyroid hormone	Regulates calcium and phosphorus levels	___
38	Glucocorticoids	Group of hormones that influence glucose metabolism	___
39	Insulin	Reduces blood glucose levels	___
40	Oestrogens	Stimulate female secondary sex characteristics	___
41	Androgens	Stimulate maturation and development of male sex organs	___

ANSWERS

LABELLING EXERCISE

Figure 6.2 The endocrine system

N.B. Figure not to scale

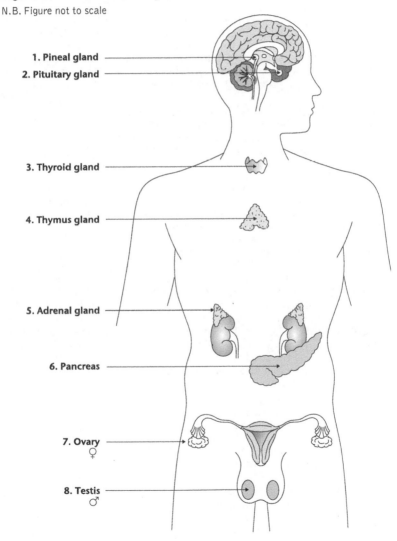

1. Pineal gland
2. Pituitary gland
3. Thyroid gland
4. Thymus gland
5. Adrenal gland
6. Pancreas
7. Ovary
   ♀
8. Testis
   ♂

1   **Pineal gland:** small cone-shaped gland situated near the centre of the brain. It is responsible for the secretion of melatonin, a hormone involved in the regulation of sleep patterns.

2   **Pituitary gland:** located at the base of the brain, the pituitary is composed of two lobes (anterior and posterior). The pituitary is linked to the hypothalamus, which stimulates the release of pituitary hormones. Although the pituitary gland is known as the master endocrine gland, both of its lobes are controlled by the hypothalamus.

The anterior pituitary synthesizes and secretes important hormones, such as corticotrophin (adrenocorticotrophic hormone, ACTH), thyroid-stimulating hormone (TSH), prolactin, growth hormone, follicle-stimulating hormone (FSH) and luteinizing hormone (LH). The posterior pituitary does not synthesize hormones but it stores and releases oxytocin and antidiuretic hormone (ADH) or vasopressin.

3   **Thyroid gland:** one of the largest endocrine glands in the body. It is found in the neck, below the thyroid cartilage and at the same approximate level as the cricoid cartilage of the larynx. The thyroid is regulated by hypothalamus and pituitary action; it controls the basal metabolic rate, synthesizes proteins and regulates sensitivity to other hormones.

The thyroid participates in these processes by producing thyroid hormones, principally thyroxine (T_4) and triiodothyronine (T_3). These hormones regulate the rate of metabolism and affect the growth and rate of function of many other systems in the body. Iodine and the amino acid tyrosine are used to form both T_3 and T_4. The thyroid also produces the hormone calcitonin, which is involved in calcium homeostasis.

Attached to the thyroid gland are the parathyroid glands. They produce parathyroid hormone (PTH) for controlling osteoclasts which facilitate bone resorption.

4   **Thymus gland:** located behind the sternum, its main function is maturation of T-cells and preventing autoimmune reactions. Its endocrine function involves the production of hormones for T-cell maturation and proliferation. There is also evidence to suggest that thymic hormones may slow the ageing process.

5   **Adrenal gland:** paired organs, one of which is located on top of each kidney. Structurally and functionally the adrenal glands are differentiated into two regions. The outer adrenal cortex makes up the bulk of the gland and surrounds the inner adrenal medulla. The adrenal cortex is divided into three zones that secrete different hormones.

6   **Pancreas:** this gland has both endocrine and exocrine functions. It is located behind and slightly below the stomach and consists of a head, body and tail. The pancreas produces endocrine and exocrine secretions. The digestive enzymes are exocrine secretions and are secreted into the duodenum via the pancreatic duct. The hormones insulin and glucagon are endocrine secretions because they are secreted directly into the blood from the pancreas.

7 **Ovary (pl. ovaries):** paired oval structures, located in the pelvic cavity of females, one on each side of the uterus and held in position by ligaments.
   The ovaries produce the female sex hormones, oestrogen and progesterone. Along with the gonadotropic hormones of the pituitary gland, these sex hormones regulate the female reproductive cycle, maintain pregnancy and prepare the mammary glands for lactation. Oestrogens and progesterone are also involved in the promotion and maintenance of female secondary sex characteristics.

8 **Testis (pl. testes):** paired structures that produce sperm and the male sex hormones, testosterone and inhibin. A system of ducts transports and stores sperm cells, assists in their maturation and conveys them to the exterior.
   The major functions of the testes are regulated by FSH and LH. In males, LH is involved in the development of interstitial cells which secrete male sex hormones. FSH prepares the seminiferous tubules (which produce sperm cells) to respond to the effects of testosterone. In the presence of FSH and testosterone, these supporting cells stimulate spermatogenic cells to undergo spermatogenesis (see *Chapter 12, Answer 35*).

TRUE OR FALSE?

9 **The pancreas has both endocrine and exocrine functions.**

The pancreas has a dual function producing hormones and pancreatic secretions with specific digestive functions. The hormones produced include glucagon and insulin (to regulate blood glucose levels), somatostatin (inhibits secretion of glucagon and insulin and slows nutrient absorption) and pancreatic polypeptide (inhibits somatostatin secretion, contraction of the gall bladder and the secretion of pancreatic digestive enzymes).
   Its main exocrine functions include the production of an alkaline pancreatic fluid which buffers acidic stomach secretions, inhibiting the action of pepsin and providing a pH more favourable to the digestive enzymes of the small intestine. The main enzymes include: pancreatic amylase, which breaks down starch; and trypsin, which digests proteins.

10 **Homeostasis of hormonal secretions is maintained only by negative feedback mechanisms.**

Normally negative feedback mechanisms maintain hormonal secretions. Blood glucose regulation is an important example of a negative feedback mechanism. Occasionally a positive feedback system can regulate hormone secretion. Usually, such amplification threatens homeostasis but in some situations can help maintain stability. During childbirth, oxytocin stimulates contractions of the uterus. These contractions stimulate additional oxytocin release and intensify the response to the initial contraction stimulus, thus establishing a positive feedback cycle. At birth, the cycle is broken by the sudden reduction of cervical distension and oxytocin secretion falls.

11 **The hypothalamus has a posterior and anterior lobe.**

The hypothalamus is a small region of the brain, below the thalamus. It controls important nervous and hormonal regulatory functions, receiving nervous input from several other regions of the brain and synthesizing at least nine different hormones, a number of which influence the production and release of several pituitary hormones. These combined nervous and endocrine functions enable the hypothalamus to regulate homeostasis of many body functions, including body temperature, control of appetite and thirst responses.

12 **Oxytocin is produced by the pineal gland.**

Oxytocin is synthesized in the hypothalamus and is stored in the posterior pituitary. Nerve impulses from the hypothalamus trigger its release into the blood. ADH is similarly produced by the hypothalamus and stored in the posterior pituitary before release.

The pineal gland is a small structure situated between the cerebral hemispheres of the brain. It produces melatonin which regulates circadian rhythms (biological clock) controlling sleep patterns. Melatonin is also linked to the onset of puberty.

13 **Body hair is produced in response to a follicle-stimulating hormone.**

The FSH is a gonadotropin produced by the anterior pituitary. In females, it initiates the development of oocytes and induces the secretion of oestrogens. In males, it stimulates the production of sperm.

The growth of body hair is stimulated by androgens. At puberty, the typical pattern of hair growth in males is associated with the secretion of significant levels of testosterone.

In females, both the ovaries and adrenal glands produce small amounts of androgens that influence the development of hair.

14 **Oestrogens are present in males.**

Oestrogens are steroid hormones present in both males and females but in significantly higher quantities in females where they are the primary female sex hormone. They are produced primarily by developing follicles in the ovaries, the corpus luteum and the placenta.

15 **Secretion of antidiuretic hormone increases urine production.**

ADH is an endocrine substance synthesized in the hypothalamus and stored in the posterior pituitary gland. Its release is stimulated by the hypothalamus.

ADH accelerates the reabsorption of water from urine in kidney tubules back into capillary blood. It also reduces the amount of water lost by the body as sweat (see Figure 6.3).

Figure 6.3 Regulation of ADH secretion and action

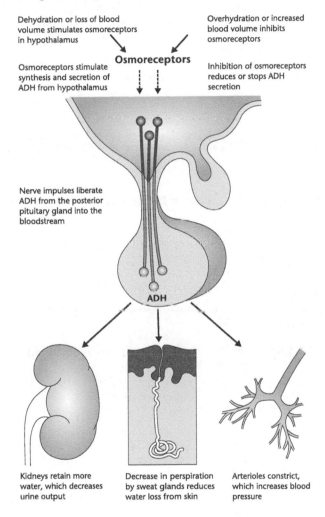

Dehydration or loss of blood volume stimulates osmoreceptors in hypothalamus

Overhydration or increased blood volume inhibits osmoreceptors

Osmoreceptors stimulate synthesis and secretion of ADH from hypothalamus

Osmoreceptors

Inhibition of osmoreceptors reduces or stops ADH secretion

Nerve impulses liberate ADH from the posterior pituitary gland into the bloodstream

ADH

Kidneys retain more water, which decreases urine output

Decrease in perspiration by sweat glands reduces water loss from skin

Arterioles constrict, which increases blood pressure

In the absence of ADH, urine output may increase in excess of tenfold normal production (1–2 L). At high concentrations ADH also causes constriction of arterioles, which increases blood pressure. The amount of ADH released by the posterior pituitary varies with blood volume and hydration status.

16 **The principal actions of thyroid hormones increase the basal metabolic rate.**

The thyroid gland secretes two major thyroid hormones: thyroxine (T_4) and tri-iodothyronine (T_3). The two hormones regulate oxygen use, basal metabolic rate, cellular metabolism and direct growth and development. Of the two thyroid hormones, T_4 is the more abundant; however, T_3 is more potent and generally considered to be the principal thyroid hormone. The thyroid gland also secretes the hormone calcitonin, which reduces blood calcium concentrations.

17 **Calcitonin is produced by the thyroid gland.**

The thyroid gland produces calcitonin in addition to the thyroid hormones, T_3 and T_4. Calcitonin decreases the concentration of calcium in blood by inhibiting bone breakdown (resorption). With reduced bone resorption, less calcium enters the blood from the bone causing a decrease of available plasma calcium. An increase in calcitonin secretion rapidly follows any increase in plasma calcium levels, causing it to return to normal.

18 **Bone resorption is facilitated by the action of the parathyroid hormone.**

The parathyroid hormone (PTH) regulates blood calcium levels in combination with calcitonin. It acts as an antagonist of calcitonin since it increases the concentration of calcium in the blood. The antagonistic effects of PTH and calcitonin are vital to normal physiology since cells and tissues are extremely sensitive to changes in blood calcium levels.

Produced by the parathyroid glands, PTH stimulates osteoclasts to increase the breakdown of bone, allowing calcium release and causing blood calcium levels to rise.

19 **The adrenal glands can be structurally and functionally differentiated into two distinct regions.**

Structurally, the adrenal glands can be differentiated into an outer cortex and an inner medulla. Adrenal cortex hormones (corticoids) differ significantly in function from those of the adrenal medulla (catecholamines).

The adrenal cortex can itself be differentiated into three distinct regions, each of which has a unique function. The outermost region (zona glomerulosa) of the adrenal cortex secretes mineralocorticoids, primarily aldosterone. The middle layer (zona fasciculata) secretes glucocorticoids including cortisol, corticosterone and cortisone. Small amounts of sex hormones are produced by the deepest region of the cortex (zona reticularis).

The adrenal medulla produces adrenaline (epinephrine) and noradrenaline (norepinephrine). Under stress, impulses received by the hypothalamus stimulate an increase in their secretion. These hormones increase blood pressure by increasing heart rate and force of contraction, in addition to constricting blood vessels. Furthermore, they cause dilation of the airways, decrease the rate of digestion, increase blood glucose, and stimulate cellular metabolism. All these responses are 'fight or flight' in nature.

20 **Glucocorticoids suppress the immune system.**

High doses of glucocorticoids depress immune responses and are prescribed to organ transplant recipients to help prevent tissue rejection by the immune system.

The glucocorticoids are also anti-inflammatory and inhibit many cells involved in inflammatory responses. Principally, they reduce the number of mast cells, thus reducing the release of histamine.

21 **The hormones produced by the adrenal medulla have major effects on heart rate and blood pressure.**

The adrenal medulla produces the catecholamines adrenaline and noradrenaline in response to sympathetic nervous impulses which affect both heart rate and blood pressure when an individual is exposed to stress. In such situations, the hypothalamus also triggers the anterior pituitary to release ACTH which stimulates the adrenal cortex to secrete glucocorticoids.

Unfortunately, during periods of prolonged stress, glucocorticoids may have harmful side effects because of their anti-inflammatory potential and capacity to cause blood vessel constriction since prolonged blood vessel constriction may lead to increased blood pressure.

22 **All endocrine glands store the hormones they produce.**

Most endocrine glands do not store their hormones but instead secrete them into the blood as required. However, the thyroid is different because it stores considerable amounts of both thyroid hormones which are released when required.

MULTIPLE CHOICE

Correct answers identified in bold italics.

23 **Melatonin is secreted by the:**

a) pancreas   ***b) pineal gland***   c) keratinocytes of the skin
d) ovaries

Melatonin is produced by the pineal gland. Its secretion is reduced in response to strong sunlight. Overproduction of melatonin has been associated with the Seasonal Affective Disorder (SAD), a type of depression relieved by exposure to repeated doses of strong artificial light. Hormones secreted by the pancreas include glucagon, insulin and somatostatin. The ovaries produce a number of female reproductive hormones. The skin produces no hormones although its appearance and structure may be significantly modified by hormonal action.

24 **The thyroid gland:**

a) is located inferior to the larynx   b) produces antidiuretic hormone   c) secretes small amounts of insulin   d) helps initiate milk production

The thyroid gland is a highly vascular structure which consists of two lobes. It is below (inferior) the larynx on either side and in front of (anterior to) the trachea. It secretes and stores two thyroid hormones which help regulate body metabolism. It is also involved in the production of calcitonin which assists in the control of blood calcium levels together with PTH.

ADH is a hormone produced by the hypothalamus and stored in the posterior pituitary gland prior to release. Insulin is secreted by the pancreas and prolactin (an anterior pituitary hormone) helps initiate milk production in the mammary glands.

25 **Glucagon is produced in the pancreas by which cells?**

a) alpha cells   b) beta cells   c) delta cells   d) F cells

The hormone glucagon, produced by pancreatic alpha cells, has the opposite effect of insulin. By raising blood glucose levels, it stimulates the liver to break down glycogen into glucose, elevating blood glucose levels. When blood glucose levels are low, a negative feedback mechanism stimulates alpha cells to release glucagon while high glucose levels reduce glucagon secretion.

Pancreatic beta cells (in the pancreatic islets) produce insulin which stimulates the liver to store glucose as glycogen which therefore removes glucose from the blood and reduces blood glucose levels. A negative feedback mechanism, sensitive to blood glucose concentrations, regulates secretion of insulin.

The delta cells of the pancreatic islets (islets of Langerhans) produce somatostatin which inhibits secretion of insulin and glucagon, as well as slowing the absorption of nutrients from the gut.

26 **Levels of which hormone are controlled by positive feedback?**

a) growth hormone　b) thyroid-stimulating hormone
c) oxytocin　d) insulin

Oxytocin stimulates contraction of smooth muscle in the pregnant uterus and is involved in the initiation and progression of childbirth. Its synthesis by the hypothalamus and secretion from the posterior pituitary are controlled by positive feedback. Increasing contraction of the uterus stimulates synthesis and secretion of more oxytocin which in turn causes an increased frequency and force of uterine contraction. The process terminates with the birth. Growth hormone, TSH and insulin levels are controlled by negative feedback mechanisms.

27 **This hormone acts on the intestines and causes increased calcium absorption:**

a) calcitonin　　*b) calcitriol*　c) thyroxine　d) pancreatic polypeptide

Calcitriol, the hormonally active form of vitamin D, is produced from a precursor (calcifediol) in the cells of the proximal tubule of the nephron in the kidneys. Calcitriol production is stimulated by a decrease in serum calcium alone or in combination with increased serum phosphate and PTH. Calcitriol production is also increased by prolactin. Calcitriol increases blood calcium levels by increasing absorption of calcium and phosphate from the GI tract and increasing renal tubular absorption of calcium and phosphate. It can also increase blood calcium through bone resorption. Calcitriol also inhibits calcitonin release, a hormone which reduces blood calcium by counteracting the effects of PTH.

28 **Which hormone is secreted in response to high blood glucose?**

a) insulin　b) glucagon　c) cortisol　d) oxytocin

Insulin is secreted by the beta cells of the pancreatic islets in response to high blood glucose concentrations and acts to reduce blood glucose concentration. An insufficient supply or inability to respond to insulin causes hyperglycaemia which is a symptom of diabetes mellitus. Glucagon, produced by pancreatic alpha cells, inhibits the action of insulin and is an antagonist of the major functions of insulin. Cortisol is a glucocorticoid secreted by the adrenal cortex which affects glucose, protein and fat metabolism. These actions help maintain blood glucose concentrations between meals. Oxytocin has no function in the regulation of blood glucose.

29 **Somatostatin is secreted by the:**

a) pancreatic delta cells　b) pancreatic polypeptide cells
c) zona fasciculata　d) posterior pituitary

Somatostatin is primarily secreted by the pancreatic delta cells located within the pancreatic islets. It inhibits the release of insulin and glucagon by the pancreatic cells and regulates various digestive functions.

Pancreatic polypeptide cells (or F cells) are located in the pancreatic islets and secrete the hormone pancreatic polypeptide which regulates the activity of pancreatic enzymes and secretions. The zona fasciculata of the adrenal cortex secretes mainly glucocorticoid hormones. The posterior pituitary does not synthesize hormones but stores oxytocin and ADH prior to release.

30 **A lack of or decrease in insulin hormone receptors on cells can result in:**

a) diabetes insipidus   b) Type 1 diabetes mellitus   *c) Type 2 diabetes mellitus*   d) juvenile diabetes

Diabetes mellitus is a group of disorders that cause an elevation in blood glucose levels (hyperglycaemia). As hyperglycaemia increases, glucose appears in the urine (glucosuria). Major symptoms include polyuria (excessive urination), polydypsia (excessive thirst) and polyphagia (excessive hunger) (see *Paramedics! Test Yourself in Pathophysiology*).

31 **Which of the following characteristics is the same for the nervous and endocrine systems?**

a) target cells affected   b) duration of actions   c) mechanism of signalling and communication   *d) none of the above*

None of these statements is common to the nervous and endocrine systems. Together the endocrine and nervous systems coordinate the function of all body systems. The nervous system controls homeostasis through nerve impulses conducted by neurones. The endocrine system releases chemical messengers that enter the blood and are distributed throughout the body.

Nervous impulses most often produce their effects within a few milliseconds, while hormones act within seconds or hours. The duration of action in response to nervous stimulation is usually brief compared to the effects of the hormonal system.

32 **The placenta produces:**

a) oestrogen   b) progesterone   c) chorionic gonadotropins
d) all of the above

These hormones are all produced by the developing placenta. Human chorionic gonadotropin (HCG) stimulates the corpus luteum in the ovary to continue the production of oestrogens and progesterone to maintain pregnancy. Placental oestrogens and progesterone help maintain pregnancy and prepare mammary glands to secrete milk.

33 **Which of the following is an example of a peptide hormone?**

a) aldosterone b) serotonin *c) gastrin* d) prostaglandins

Gastrin is a peptide hormone produced by the G cells in the stomach. It stimulates secretion of large amounts of gastric juice. It also causes contraction of the lower oesophageal sphincter, increases stomach motility and relaxes the pyloric and ileocaecal sphincters. Gastrin secretion is inhibited when the pH of gastric juice drops below 2.0 and is stimulated when it rises, maintaining an optimum pH for the activity of the enzyme pepsin which denatures proteins in the stomach and inhibits microbial growth.

Serotonin (5-HT) is a hormone secreted from enterochromaffin (EC) cells of the GI tract where it controls intestinal movement. It is stored in blood platelets. When the platelets bind to a clot, they release serotonin which acts as a vasoconstrictor, helping to regulate haemostasis and blood clotting. Serotonin also is a growth factor for some types of cells, which suggests a role in wound healing.

The prostaglandins are membrane-bound lipids which are released in small amounts from the many tissues in which they are formed. They demonstrate a variety of strong physiological effects, such as regulating smooth muscle contraction and relaxation.

 MATCH THE TERMS

Hormone	*Response*	*Gland*
34 Trophic hormones	Exert their effects on activities of pituitary gland. This gland also secretes two other hormones (ADH and oxytocin) that are stored in the pituitary for later release	**F**. hypothalmus
35 Antidiuretic hormone	Enhances water conservation by the kidneys	**C**. pituitary
36 Thyroxine	Regulates the metabolic rate and energy production	**A**. thyroid
37 Parathyroid hormone	Regulates calcium and phosphorus levels	**E**. parathyroids
38 Glucocorticoids	Group of hormones that influence glucose metabolism	**B**. adrenals

	Hormone	*Response*	*Gland*
39	Insulin	Reduces blood glucose levels	**G.** pancreas
40	Oestrogens	Stimulate female secondary sex characteristics	**D.** ovaries
41	Androgens	Stimulate maturation and development of male sex organs	**H.** testes

7 The cardiovascular system

INTRODUCTION

The cardiovascular (circulatory) system pumps and channels blood containing glucose and oxygen around the body. A pump system is necessary because diffusion is only efficient over short distances. The human circulatory system is a double circulatory system consisting of the pulmonary and systemic circulations.

Blood vessels transport blood around the body. Their structure varies according to their slightly different functions. Arteries are elastic and have a thick muscular wall with a relatively narrow lumen. Veins have a larger lumen than arteries and a relatively thin, less elastic wall. Capillaries are only one cell thick and are the site of nutrient and gas exchange between blood and tissues. Oxygenated blood travels away from the heart under high pressure in the arteries. Deoxygenated blood is carried towards the heart in veins under much lower pressure; valves exist in veins to prevent back-flow of blood.

Blood is 55 per cent plasma, 45 per cent red blood cells and less than 1 per cent white blood cells and platelets. As blood flows through the tissues, dissolved substances are exchanged between the cells and blood in capillaries. As a paramedic, it is essential to understand the structure and function of the cardiovascular system since it is the main transportation system for the body, facilitating the movement of essential nutrients and gases throughout all of the body's systems.

Useful resources

Paramedics! Test Yourself in Pathophysiology

Chapter 7

Anatomy and Physiology (8th edition)

Chapters 16, 17 and 18

Ross and Wilson's Anatomy and Physiology in Health and Illness (11th edition)

Chapters 4 and 5

Interactive tutorial on the cardiovascular system:

http://www.getbodysmart.com/ap/circulatorysystem/menu/menu.html

---

LABELLING EXERCISE

1–11 Identify the regions of the human heart in Figure 7.1, using the terms provided in the box below.

aorta	bicuspid valve
sinoatrial node	tricuspid valve
aortic valve	Purkinje fibres
pulmonary valve	bundle of His
left ventricle	bundle branches
atrioventricular node	

Figure 7.1 The human heart

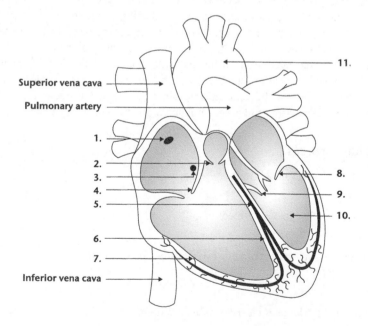

12–15 Identify the main vessels of the circulatory system in Figure 7.2, using the terms provided on p. 104.

Figure 7.2 The main vessels of the circulatory system

N.B. Figure not to scale

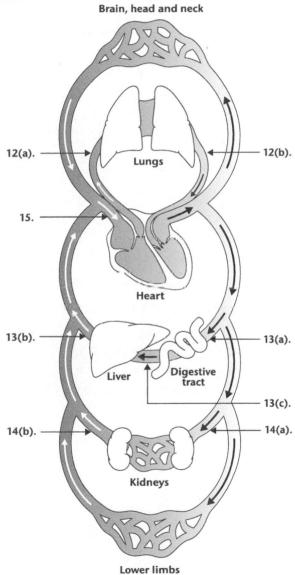

renal artery

pulmonary veins

hepatic vein

pulmonary arteries

renal vein

venae cavae

hepatic portal vein

hepatic artery

TRUE OR FALSE?

Are the following statements true or false?

16 The left side of the heart is more muscular than the right.

17 The cardiac cycle is a sequence of events occurring in one minute.

18 There are two different types of clotting pathway.

19 The bundle of His is not involved in pumping blood since it has no contractile proteins.

20 The sinoatrial (SA) node is located in the left atrium.

21 Systole refers to relaxation of the heart chambers.

22 Cardiac muscle cells are best described as voluntary striated muscle tissue.

23 Cardiac muscle is myogenic.

24 All arteries carry oxygenated blood and all veins transport deoxygenated blood.

MULTIPLE CHOICE

Identify one correct answer for each of the following.

25 Name the arteries and veins facilitating blood flow to and from the lower limbs:

a) brachial
b) mesenteric
c) femoral
d) carotid

26 Blood flows out of the ventricles when:

a) the atrioventricular valves are open
b) the semi-lunar valves are open
c) the bicuspid valves are open
d) the mitral valves are closed

27 Cardiac output is the amount of blood pumped by:

a) 1 ventricle in 1 minute
b) 1 atrium in 1 minute
c) both ventricles in 1 minute
d) both atria in 1 minute

28 Blood from the heart flows through the blood vessels in which order?

a) veins, arterioles, capillaries, venules, arteries
b) capillaries, arterioles, arteries, venules, veins
c) arteries, arterioles, capillaries, venules, veins
d) arteries, capillaries, arterioles, venules, veins

29 Blood pressure is highest when leaving which heart chamber?

a) right atrium

b) right ventricle

c) left atrium

d) left ventricle

30 Blood pressure is usually expressed as:

a) diastolic pressure over systolic pressure

b) systolic pressure over diastolic pressure

c) diastolic pressure over pulse pressure

d) pulse pressure over diastolic pressure

31 Which of the following statements obeys Starling's law?

a) increasing venous return, increases stroke volume

b) increasing blood pressure, increases stroke volume

c) increasing stroke volume, increases pulse rate

d) increasing pulse rate, increases cardiac output

32 Divisions of the systemic circulation are usually named according to:

a) cells that make them up

b) bones they supply

c) tissues/organs they supply

d) people who first identified them

FILL IN THE BLANKS

Fill in the blanks in each statement using the options in the box below.
Not all of them are required, so choose carefully!

120/80	160/90
single	semi-lunar
electrocardiogram (ECG) trace	parasympathetic
atrioventricular valves	sympathetic
systemic	chemoreceptors
baroreceptors	cardiac output
CT scan	action potential
double	

---

33 The human heart is referred to as a _____ circulation.

34 Electrical impulses of the heart can be detected on the surface of the skin by means of an _____ _____.

35 The normal resting blood pressure for a healthy adult is expressed as approximately _____ / _____ mmHg.

36 The two main factors for determining blood pressure are peripheral resistance and _____ _____.

37 _____ nerves increase heart rate and force of heart beat.

38 During each cardiac cycle two heart sounds are heard. These are due to the closing of _____ _____ followed by the closing of _____ valves.

39 _____ are sensory receptors located in the blood vessels that detect the pressure of blood flowing through them.

40 An _____ _____ is a short-lasting event in which the electrical membrane potential of a cell rapidly rises and falls.

41 Label the action potential from the options in the box below.

calcium channel opening   repolarization of potassium ions

refractory period   sodium channel inactivation

influx of sodium ions

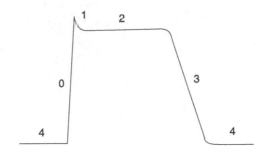

ANSWERS

 LABELLING EXERCISE

Figure 7.3 The human heart

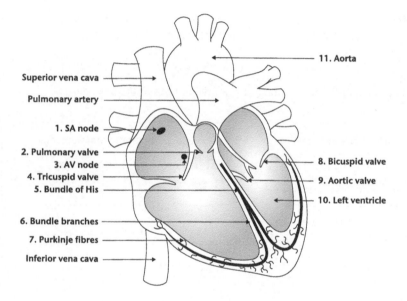

11. Aorta

Superior vena cava

Pulmonary artery

1. SA node

2. Pulmonary valve
3. AV node
4. Tricuspid valve
5. Bundle of His

8. Bicuspid valve
9. Aortic valve
10. Left ventricle

6. Bundle branches
7. Purkinje fibres
Inferior vena cava

Figure 7.4 The main vessels of the circulatory system

N.B. Figure not to scale

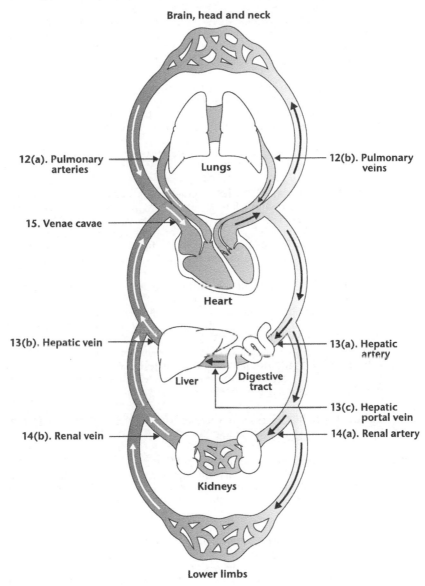

Brain, head and neck

12(a). Pulmonary arteries

12(b). Pulmonary veins

Lungs

15. Venae cavae

Heart

13(b). Hepatic vein

13(a). Hepatic artery

Liver

Digestive tract

13(c). Hepatic portal vein

14(b). Renal vein

14(a). Renal artery

Kidneys

Lower limbs

1 | *Sinoatrial (SA) node:* a region of electrically excitable tissue located in the right atrium near the superior vena cava. Every heartbeat is initiated here when the SA node (also known as the pacemaker) generates an electrical signal that spreads over both atria causing them to contract. This sets the rhythm of the heartbeat. The right atrium always contracts just before the left atrium because of the short time taken for the signal to travel through the tissue to the atrioventicular node.

2 | *Pulmonary (semi-lunar) valve:* when the right ventricle contracts, it forces blood out of the chamber into the pulmonary artery. The blood passes out of the right ventricle when the semi-lunar valve opens. The valve closes, preventing blood flowing back into the heart after the ventricle contracts. From the pulmonary artery, blood travels to the lungs to become oxygenated. (If in doubt, use 'semi-lunar valve' because it describes the valves on either side of the heart.)

3 | *Atrioventricular (AV) node:* another mass of tissue located in the lower part of the right atrium in the septal wall at the junction with the right ventricle. The electrical signal moves from the SA node to the AV node. The AV node slows the speed of the electrical signal providing a slight delay which allows the atria to fully contract and gives the ventricles time to fill up before expelling blood.

4 | *Tricuspid (atrioventricular) valve:* as the right atrium contracts, this valve opens allowing blood to flow into the right ventricle. As blood enters the chamber, pressure increases. The tricuspid valve closes to prevent back-flow of blood into the atrium. *Think right side = tricuspid valve.* (If in doubt, use 'AV valve'.)

5 | *Bundle of His (atrioventricular bundle):* the signal from the AV node then passes along the bundle of His which are modified muscle fibres in the septum of the heart.

6 | *Bundle branches:* these separate out over the right and left sides of the heart, carrying the electrical signal across the ventricles.

7 | *Purkinje fibres:* the distal portion of the left and right bundle branches from the bundle of His. They surround the ventricles from the endocardium to the myocardium carrying the electrical impulse to the apex (pointed end) of each ventricle, causing the ventricle muscles to contract from the apex upwards.

8 | *Bicuspid (mitral, atrioventricular) valve:* oxygenated blood returns from the lungs to the left atrium via the pulmonary vein. After the left atrium fills with blood, it contracts and the bicuspid valve opens, allowing blood to flow into the left ventricle. The valve then closes, preventing back-flow of blood into the atrium. (If in doubt, use 'AV valve' to prevent confusion between valves of the left and right atria.)

9 *Aortic (semi-lunar) valve:* when the pressure in the left ventricle increases, it forces open the semi-lunar valve allowing the blood to flow out of the left ventricle. The valve closes again when the pressure drops, preventing back-flow of blood into the ventricle. (If in doubt, use 'semi-lunar valve' because it describes the valves on either side of the heart.)

10 *Left ventricle:* receives oxygenated blood from the left atrium. The left ventricle has more muscle around it since it pumps blood all around the body in the systemic circulation. As the left ventricle fills with blood, the pressure in the ventricle increases, generating enough pressure to force the blood into the systemic circulation.

11 *Aorta:* when blood leaves the left ventricle it enters the aorta, which is the main artery carrying oxygenated blood away from the heart. It is a very elastic artery allowing the aorta to accommodate surges of blood entering the aorta at high pressure from the left ventricle. As the ventricle relaxes, the aorta recoils and pushes the oxygenated blood onwards into the systemic circulation at high pressure.

12 **(a)** *Pulmonary arteries:* deoxygenated blood leaves the right side of the heart via the two pulmonary arteries and travels to the lungs where it becomes oxygenated.

12 **(b)** *Pulmonary veins:* after being oxygenated, blood travels from the lungs back to the heart via the four pulmonary veins, which enter the left side of the heart.

13 **(a)** *Hepatic artery:* delivers oxygenated blood to the liver where oxygen demands are very high.

13 **(b)** *Hepatic veins:* the blood is then drained in the gastrointestinal tract and spleen where it enters the hepatic veins and is taken back to the heart.

13 **(c)** *Hepatic portal vein:* takes blood from the capillary beds in the digestive tract and transports it to the capillary beds in the liver (because portal veins carry blood between capillary beds).

14 **(a)** *Renal artery:* carries oxygenated blood and high concentrations of solutes such as urea, creatinine and sodium ions (Na^+) to the kidneys.

14 **(b)** *Renal vein:* returns deoxygenated blood from the kidneys to the general circulation. After leaving the kidneys, via the renal vein, the blood will have a lower concentration of urea and creatinine and the blood pH, along with the ratio of sodium to potassium ions, will be regulated.

15 *Venae cavae (pl. for inferior and superior vena cava):* deoxygenated blood returns to the right side of the heart via the venae cavae. The superior vena cava returns deoxygenated blood primarily from the head and forelimbs. The inferior vena cava returns deoxygenated blood from the rest of the body. The venous return to the heart initiates the expansion in the venae cavae that triggers the SA node to fire the electrical impulse which generates a heartbeat.

TRUE OR FALSE?

16 **The left side of the heart is more muscular than the right.** ✓

The left ventricle has a thicker muscular wall than the right ventricle because it needs to generate a higher pressure in the blood as it leaves the left side of the heart entering the aorta to be distributed throughout the systemic circulation. Therefore, the pressure of the muscle contraction is greater around the left ventricle than the other chambers.

17 **The cardiac cycle is a sequence of events occurring in one minute.** ✖

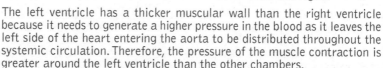

Events of the cardiac cycle occur during one heartbeat and take approximately 0.8 seconds to complete each cycle (0.1 seconds for atrial systole, 0.3 seconds for ventricular systole, and 0.4 seconds for total cardiac diastole). It is usually described as starting when blood enters the atria. The atria contract (0.1 seconds), pushing blood into the ventricles through the atrioventricular (bicuspid or tricuspid) valves. The ventricles contract (0.3 seconds), forcing blood into the aorta and pulmonary artery via the semi-lunar valves. Blood leaves the heart via the arteries and due to the absence of electrical activity, or total cardiac diastole (0.4 seconds), the atria begin to fill with blood again and the cardiac cycle begins again.

18 **There are two different types of clotting pathway.** ✖

There are three different clotting (coagulation) pathways: the extrinsic pathway, the intrinsic pathway, and the common pathway. The extrinsic pathway is activated by damaged tissue while the intrinsic pathway is activated when plasma comes in contact with a damaged region of a vessel. The extrinsic pathway has fewer steps than the intrinsic route and occurs very quickly (usually within a matter of seconds, depending on the injury) resulting in the mixing of factor X, factor V and Ca^{2+} to produce the enzyme prothrombinase. The intrinsic pathway is more complex than the extrinsic pathway since it requires activation of factor XII, factor X, platelets and Ca^{2+} to form prothrombinase. Irrespective of which pathway initiates coagulation, the final stages of the process occurs via the common pathway. During the common pathway, prothrombin is converted to thrombin through the action of prothrombinase in the presence of Ca^{2+} and fibrinogen is processed into fibrin threads. Effective clotting requires several coagulation factors (identified by roman numerals).

19 **The bundle of His is not involved in pumping blood since it has no contractile proteins.**

The bundle of His (AV bundle) is a collection of specialized heart muscle cells involved in electrical conduction. They transmit the electrical impulses from the AV node. Therefore, they are not involved in physically moving blood around the heart; rather, they are involved in generating the heartbeat in the muscle of the heart.

20 **The sinoatrial (SA) node is located in the left atrium.**

The SA node is the impulse-generating (pacemaker) tissue located in the right atrium, near the entrance of the superior vena cava. Although all the heart's cells possess the ability to generate the electrical impulses (action potentials) that trigger cardiac contraction, the SA node normally initiates it. When a patient's SA node is not functioning properly, they may be fitted with an artificial pacemaker which mimics the action of the SA node and maintains a steady heart rate.

In the absence of extrinsic neural and hormonal control, cells in the SA node will spontaneously create action potentials at about 60–100 beats per minute. Nerves in the SA node are supplied by the parasympathetic and sympathetic nervous systems. Parasympathetic nerves only change heart rate; they cannot change the force of contraction because they only innervate the SA node and AV node. However, sympathetic nerves can increase the force of heart muscle contraction because they innervate the atria and ventricles as well as the SA and AV nodes.

21 **Systole refers to relaxation of the heart chambers.**

Systole is contraction of the heart chambers. It describes a phase of the cardiac cycle where the myocardium is contracting in response to the electrical stimulus initiated by the SA node. During systole, pressure is being generated within the heart chambers driving the blood flow. All four heart chambers undergo systole and diastole in a timed fashion so that blood is propelled forward through the cardiovascular system, preventing back-flow of blood in the heart vessels.

22 **Cardiac muscle cells are best described as voluntary striated muscle tissue.**

Cardiac muscle is involuntary, striated muscle. It is unique tissue, structurally similar to both smooth and skeletal muscle. It has a very similar structure to skeletal muscle with visible cross-striations and contractile proteins. This is important, since the heart muscle contraction needs to be strong and efficient with sufficient force to propel blood throughout the systemic circulation. The extensive branching existing between cardiac myocytes at the intercalated discs is unique to cardiac muscle cells. The branching enables the myocytes to develop force in many different directions, facilitating the efficient expulsion of blood from the cardiac chambers.

23 Cardiac muscle is myogenic.

Cardiac muscle is very specialized as it is the only type of muscle that has an internal rhythm, it is myogenic. This means that it can naturally contract and relax without receiving electrical impulses from nerves. When cardiac muscle cells are placed next to another, they will beat in unison, this is functional syncytium. It is this pattern of contractions that maintains a regular heartbeat.

24 All arteries carry oxygenated blood and all veins transport deoxygenated blood.

The pulmonary artery is the only artery in the body that carries deoxygenated blood, while the pulmonary vein is the only vein transporting oxygenated blood. These two vessels make up the pulmonary circuit of the double circulatory system. It is correct to say that arteries always carry blood away (*hint:* <u>*a*</u>*rteries* = <u>*a*</u>*way*) from the heart and veins always carry blood towards the heart. Arteries are better equipped to deal with high internal flow pressures, as they have a dominant layer of smooth muscle and elastic tissue (tunica media). Veins have the same smooth muscle component, although it is not as thick as that in arteries. Instead, veins rely on the contraction of skeletal muscles to move flow along, together with valves (mostly found in the limbs) that prevent back-flow.

 MULTIPLE CHOICE

Correct answers identified in bold italics.

25 Name the arteries and veins facilitating blood flow to and from the lower limbs:

a) brachial   b) mesenteric   *c) femoral*   d) carotid

The femoral artery supplies oxygenated blood to the lower limbs. The vessel runs along the inner groin. The femoral vein returns deoxygenated blood from the lower limbs to general circulation. Blood in the femoral vein may contain high concentrations of lactic acid after exercise. The vessels contain numerous valves to reduce blood pooling in the lower limbs. The brachial artery and vein service the upper limbs (arms). The mesenteric artery carries oxygenated blood to the gastrointestinal (GI) tract. Deoxygenated blood is removed from the GI tract via the hepatic vein. The carotid artery transports oxygenated blood to the head and neck; within its walls there are baroreceptors for detecting change in blood pressure and chemoreceptors for detecting changes in carbon dioxide and oxygen in the blood.

26 | Blood flows out of the ventricles when:

a) the atrioventricular valves are open   *b) the semi-lunar valves are open*   c) the bicuspid valves are open   d) the mitral valves are closed

The semi-lunar valves are located between the ventricles and the arteries leaving both sides of the heart. The valves act as gates which open when pressure in the ventricles exceeds the pressure in the artery. After blood leaves the ventricles and enters the arteries, the valves will close, preventing back-flow of blood into the ventricles.

27 | Cardiac output is the amount of blood pumped by:

a) 1 ventricle in 1 minute   b) 1 atrium in 1 minute   c) both ventricles in 1 minute   d) both atria in 1 minute

Cardiac output relates heart rate (HR) and stroke volume (SV). It is calculated by multiplying the heart rate by the stroke volume. Stroke volume refers to the amount of blood expelled from the heart in one minute. Cardiac output is related to body mass. The cardiac output of a resting 70 kg male is 5–6 litres (or 5000–6000 millilitres) per minute. In normal resting conditions, venous return (that is, return of blood to the heart) is equal to cardiac output. During exercise, the cardiac output can increase to as high as 30 litres per minute. This is due to the increased heart rate which is caused by an increase in the stroke volume.

28 | Blood from the heart flows through the blood vessels in which order?

a) veins, arterioles, capillaries, venules, arteries   b) capillaries, arterioles, arteries, venules, veins   *c) arteries, arterioles, capillaries, venules, veins*   d) arteries, capillaries, arterioles, venules, veins

As blood flows through the vascular system, it travels through five types of blood vessels. The structure of each vessel differs according to its function and the pressure exerted by the volume of blood. Arteries have thick, muscular walls to accommodate the flow of blood at high speed and pressure. Around the heart the arteries are elastic to accommodate the rise in pressure during systole. Arteries are distinguished from arterioles by having three or more layers of smooth muscle in their walls. Arterioles have thinner walls than arteries; they dilate and constrict to control the flow of blood to the capillaries. Capillary walls are one cell thick and selectively permeable to allow exchange of gases, nutrients and waste between cells and blood. The venules receive blood from the capillaries so their walls are slightly thicker than capillaries but thinner than arterioles, the venules possess valves that prevent the back-flow of blood. From the venules, blood then flows into the veins which carry it back to the heart. Veins are less muscular than arteries but have larger diameter so blood is under less pressure in veins than in arteries. Veins do not have valves but smooth muscles contract to prevent back-flow. The difference between veins and arteries is the direction they carry blood in rather than the oxygen content of the blood. (*Remember: arteries away from the heart; veins towards heart.*)

29 | Blood pressure is highest when leaving which heart chamber?

a) right atrium   b) right ventricle   c) left atrium   *d) left ventricle*

Blood pressure (BP) is defined as the force exerted on the walls of the arteries as the blood moves through them. Every heartbeat expels a quantity of blood out of the heart into the blood vessels. More muscle surrounds the left side of the heart than the right, therefore when the chambers of the left side contract, blood is forced out with much greater pressure than when it leaves the right side. This is because when blood exits the left side of the heart, it travels around the entire body in the systemic circulation whereas blood leaving the right side of the heart is only travelling to the lungs and back via the pulmonary circulation. Since the left ventricle is the last chamber the blood occupies before entering the systemic circulation, it must expel the blood with enough pressure to travel around the whole body.

30 | Blood pressure is usually expressed as:

a) diastolic pressure over systolic pressure   *b) systolic pressure over diastolic pressure*   c) diastolic pressure over pulse pressure   d) pulse pressure over diastolic pressure

Blood pressure is one of the principal vital signs. During each heartbeat, BP varies between a maximum (systolic) and a minimum (diastolic) pressure. BP is highest when leaving the left ventricle at the height of the heart's contraction; this is systolic blood pressure and is normally around 120 mmHg. BP is lowest when the heart is relaxing during diastole. Diastolic pressure is usually around 80 mmHg. The conducting arteries around the heart are elastic and therefore can stretch and recoil to accommodate the pressure of the blood leaving the heart. As the blood passes through the conducting arteries to the distributing arteries, the pressure drops because the elastic muscles of the conducting arteries absorb some of the initial pressure. The pressure continues to fall as the blood is dispersed through the distributing arteries and into the arterioles where BP is usually around 40 mmHg.

31 | Which of the following statements obeys Starling's law?

a) increasing venous return, increases stroke volume   b) increasing blood pressure, increases stroke volume   c) increasing stroke volume, increases pulse rate   d) increasing pulse rate, increases cardiac output

The volume of blood entering the heart should always equal the volume of blood leaving the heart. Starling's law (Frank-Starling mechanism) states that the heart is able to change the force of its contraction (and hence its stroke volume) in response to changes in venous return. The greater the volume of blood entering the heart during diastole, the greater the volume of blood expelled during systolic contraction (stroke volume). As blood enters the left ventricle, it stretches the ventricle wall. The more the wall is stretched, the more the ventricle muscles contract and more blood is pumped out of the heart. This stimulates stretch receptors which trigger a reflex increase in heart rate. This allows cardiac output to be synchronized with venous return and arterial blood supply.

32 **Divisions of the systemic circulation are usually named according to:**

a) cells that make them up   b) bones they supply   *c) tissues/organs they supply*   d) people who first identified them

There are many divisions of the systemic circulation, usually named after the organs, tissues or system they serve. Each organ or system has an artery bringing oxygenated blood and nutrients and a vein removing deoxygenated blood and waste. The systemic circuit serves the rest of the body (but not the lungs). The pulmonary circuit supplies only the lungs.

 # FILL IN THE BLANKS

33 **The human heart is referred to as a *double* circulation.**

Double circulation refers to the heart acting like two separate pumps with blood passing through the heart twice before being pumped around the body. Together the pulmonary and systemic circuits comprise the double circulatory system yet the two circuits are completely independent of each other.

34 **Electrical impulses of the heart can be detected on the surface of the skin by means of an *electrocardiogram (ECG) trace.***

Electrical pulses within cardiac tissue stimulate the relaxation and contraction of the myogenic muscle around the heart. These electrical impulses can be detected on the surface of the skin and visualized as an electrocardiogram (ECG) trace which measures electrical activity in the heart over a period of time. A typical ECG tracing of a normal heartbeat (cardiac cycle) consists of a P wave, a QRS complex and a T wave (see Figure 7.5). In 50–75 per cent of traces, a small U wave is observed; it is typically small and follows the T wave and is known to represent papillary or repolarization of Purkinje fibres. The baseline voltage of the ECG is known as the isoelectric line which is measured as the portion of the tracing following the T wave and preceding the next P wave.

- *P wave*: During normal atrial contraction, the SA node sends an electrical signal to the AV node which spreads from the right atrium to the left atrium. The time taken for the electrical signal to go from the SA to the AV is called the PR interval, and takes between 120 and 200 ms (0.12–0.20 seconds).

- *QRS complex:* Represents contraction of both ventricles – this should take no longer than 120 ms (0.12 seconds).

Figure 7.5 Electrocardiogram trace

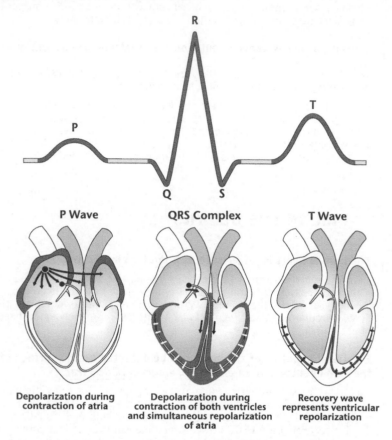

P Wave

Depolarization during
contraction of atria

QRS Complex

Depolarization during
contraction of both ventricles
and simultaneous repolarization
of atria

T Wave

Recovery wave
represents ventricular
repolarization

- *T wave:* Relaxation and recovery (repolarization). The interval from the beginning of the QRS complex to the apex (highest point) of the T wave is referred to as the absolute refractory period; during this phase no action potential can be generated. The last half of the T wave is referred to as the relative refractory period.

35 **The normal resting blood pressure for a healthy adult is expressed as approximately *120/80* mmHg.**

Arterial blood pressure is measured using a sphygmomanometer, which historically used the height of a column of mercury to reflect the circulating pressure. BP values are reported in millimetres of mercury (mmHg), although most devices do not use mercury. The term *blood pressure* usually refers to the pressure measured inside the elbow in the brachial artery of the upper arm. This is the upper arm's major blood vessel carrying blood away from the heart. BP is commonly measured by the auscultatory method. A cuff is fitted around the upper arm and inflated by repeatedly squeezing a rubber bulb until the artery is completely occluded (squeezed closed). While listening with the stethoscope to the brachial artery at the elbow, the pressure in the cuff is slowly released. When blood just starts to flow again in the artery, the turbulent flow creates a 'whooshing' or pounding sound. The pressure at which this sound is first heard is recorded as the systolic BP. The cuff pressure is further released until no sound can be heard – this is the diastolic arterial pressure. For a healthy resting adult, the normal BP is said to be '120 over 80 millimetres of mercury'. BP values vary with age, sex, exercise, sleep and emotion. Age-for-age, BP is normally lower for females than males. In children, BP measurements are lower than adults and vary with height. In the older adult, BP tends to be higher due to the decrease in elasticity of the blood vessels that accompanies the ageing process. (See *Paramedics! Test Yourself in Pathophysiology* for details on abnormal blood pressures.)

36 **The two main factors for determining blood pressure are peripheral resistance and *cardiac output*.**

Cardiac output indicates how well the heart is performing its function of transporting blood to the cells. Cardiac output is primarily regulated by the cells' demand for oxygen. If the cells are working hard, they demand more oxygen, the cardiac output is raised to increase the oxygen supply to cells, while at rest when the cellular oxygen demand is low, the cardiac output is said to be baseline. Cardiac output is regulated not only by the heart rate, but also by the circulatory vessels as they relax and contract, increasing and decreasing the resistance to flow, that is, the peripheral resistance.

37 **_Sympathetic_ nerves increase heart rate and force of heart beat.**

Along with electrical stimulation, certain nervous reflexes regulate heart rate. Sensory nerves in the wall of the aorta and atria respond to stretching in these vessels as a result of increased venous return. In the aorta, these receptors (baroreceptors) detect increasing blood pressure. When blood pressure is high, they trigger a reflex, slowing the heart and reducing cardiac output and blood pressure. When blood pressure is low, the heart rate increases, as in shock. There are similar pressure receptors in the atria. Arteries and arterioles are regulated by the autonomic nervous system and the vasomotor centre in the medulla oblongata. During sympathetic nervous control, the vasomotor centre stimulates overall vasoconstriction causing an increase in systemic blood pressure.

38 **During each cardiac cycle two heart sounds are heard. These are due to the closing of _atrioventricular valves_ followed by the closing of _semi-lunar_ valves.**

The two heart sounds ('lub' and 'dub') can be heard when a stethoscope is applied to the chest wall. The first sound ('lub') is longer, softer and lower pitched. It is heard when the atrioventricular valves close at the beginning of ventricular systole. The second sound ('dub') is shorter, sharper and higher pitched. It is heard when the semi-lunar valves close during ventricular diastole. These sounds are repeated every cardiac cycle and are heard about 70 times per minute in a healthy adult.

39 ***Baroreceptors*** **are sensory receptors located in the blood vessels that detect the pressure of blood flowing through them.**

Baroreceptors detect the stretching of the blood vessel walls caused by blood flowing through them and send a signal to the CNS in response to the stretch. The CNS responds to increases or decreases in total peripheral resistance and cardiac output. When there is a change in the normal arterial BP, the baroreceptors respond immediately through a negative feedback system called the baroreceptor reflex. This helps return the pressure to normal and maintain a stable BP. If BP becomes elevated, the baroreceptors are stimulated and the reflex decreases BP. When BP drops, the reflex is suppressed allowing BP to rise. However, the baroreceptors only respond to short-term changes in BP. Long-term control of BP involves regulation of blood volume by the kidneys and the renin-angiotensin-aldosterone system (see *Chapter 10, Answer 22*).

40 **An *action potential* is a short-lasting event in which the electrical membrane potential of a cell rapidly rises and falls.**

When a channel opens, it allows an inward flow of sodium ions, which changes the electrochemical gradient of a cell. This then produces a further rise in the membrane potential of a cell, causing more channels to open, and producing a greater electric current. This continues until all available ion channels are open, resulting in a positive rise in the membrane potential. The sodium ions then cause the polarity of the plasma membrane to reverse, and the ion channels 'turn off'. As the sodium channels close, sodium cannot enter the neuron, and so are transported out the plasma membrane. Potassium channels are then activated, which return the electrochemical gradient to a resting state. After the action potential, a refractory period happens, which prevents an action potential travelling back the way it just came.

41 **Label the action potential**

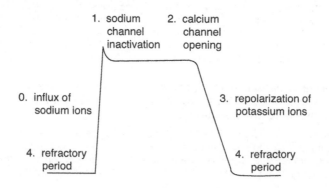

1. sodium channel inactivation
2. calcium channel opening
0. influx of sodium ions
3. repolarization of potassium ions
4. refractory period
4. refractory period

8 The respiratory system

INTRODUCTION

Cells continually use oxygen (O_2) for the metabolic reactions that release energy from nutrient molecules and generate energy as adenosine triphosphate (ATP). At the same time these reactions release carbon dioxide (CO_2). Since an excessive amount of CO_2 produces acidity that is toxic to cells, CO_2 must be eliminated quickly and efficiently. The cardiovascular system and the respiratory system cooperate to supply cells and tissues with O_2 and eliminate CO_2.

The respiratory system facilitates gas exchange, intake of O_2 and elimination of CO_2, whereas the cardiovascular system transports the gases in the blood between the lungs and body cells. Failure of either system has the same effect on the body: disruption of homeostasis and rapid death of cells from oxygen starvation and accumulation of waste products.

In addition to its role in gas exchange, the respiratory system also contains receptors for the sense of smell, filters inspired air, produces sounds and assists in the elimination of waste. Since paramedics are often responsible not only for the ventilation of patients, but for full airway care and protection, they should understand how the lungs function in gas exchange and know how their structure is optimized for this purpose.

Useful resources

Paramedics! Test Yourself in Pathophysiology

Chapter 8

Anatomy and Physiology (8th edition)

Chapter 20

Interactive tutorial on the respiratory system:

http://www.getbodysmart.com/ap/respiratorysystem/menu/menu.html

---

LABELLING EXERCISE

1–8 Identify the regions of the human respiratory system in Figure 8.1, using the options provided in the box below.

upper respiratory tract	left primary bronchus
bronchiole	lower respiratory tract
hilum	external nares
alveolus	trachea

Figure 8.1 The human respiratory system

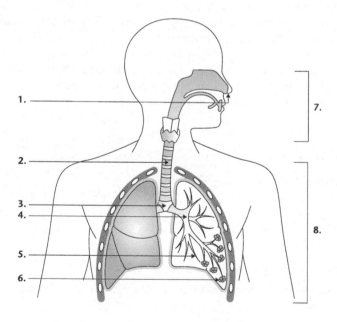

TRUE OR FALSE?

Are the following statements true or false?

9 Free movement of a gas always proceeds from a region of high pressure to a region of lower pressure.

10 Pulmonary ventilation is the exchange of gases between the atmosphere and alveoli.

11 The pO_2 in alveolar air is approximately 13.3 kPa.

12 The right primary bronchus is more horizontal than the left.

13 The basic respiration rhythm is controlled by parts of the nervous system in the medulla oblongata and pons.

14 Lung capacity typically increases with age.

15 A peak flow meter measures ability to inhale air.

16 The respiratory system is involved in regulating blood pH.

17 The epiglottis serves to protect the lower respiratory tract from foreign matter.

18 The epiglottic vallecula is part of the lower respiratory tract.

19 The cricothyroid membrane is constructed of fibrolastic tissue.

MULTIPLE CHOICE

Identify one correct answer for each of the following.

20 Hypoxia is a tissue-level deficiency of:

a) oxygen

b) carbon dioxide

c) carbon monoxide

d) water

21 The pharynx is divided into how many sub-divisions?

a) 2

b) 3

c) 4

d) 5

22 The posterior thoracic cage is composed of how many pairs of ribs?

a) 8

b) 10

c) 12

d) 14

23 The precise site of gas exchange within the lungs occurs at:

a) the nasal mucosa

b) the bronchi

c) the bronchioles

d) the alveoli

24 The wandering phagocytes found in the alveoli are called:

  a) alveolar macrophages
  b) pulmonary cells
  c) goblet cells
  d) chalice cells

25 Carbon dioxide is carried from body tissues in what form?

  a) dissolved CO_2
  b) oxyhaemoglobin
  c) carbaminohaemoglobin
  d) bicarbonate ions

26 Lung volumes exchanged during breathing are measured using:

  a) a stethoscope
  b) a spirometer
  c) a sphygmomanometer
  d) a bronchoscope

27 The total lung capacity in a healthy adult male is approximately:

  a) 500 mL
  b) 5000 mL
  c) 6000 mL
  d) 6000 L

FILL IN THE BLANKS

Fill in the blanks in each statement using the options in the box below.
Not all of them are required, so choose carefully!

bronchoscopy	6
heart	larynx
pneumotaxic	compliance
laparoscopy	reluctance
lungs	surfactant
abdominal	haemoglobin
external	18
internal	thoracic
12	

28 Blood travels from the pulmonary arteries to the _____.

29 Most of the oxygen transported in blood is bound to _____.

30 _____ is the visual examination of the bronchus through a bronchoscope.

31 The vocal cords are contained within the _____.

32 The production of _____ by type II alveolar cells reduces surface tension within the lungs.

33 The lungs are contained within the _____ cavity.

34 The exchange of gases between alveoli and their surrounding capillaries is called _____ respiration.

35 Healthy adults take approximately __ breaths per minute.

36 _____ is the ease with which the lungs and thoracic wall expand.

37 The respiratory centre consists of a medullary rhythmicity area, _____ area and an apneustic area.

ANSWERS

LABELLING EXERCISE

Figure 8.2 The human respiratory system

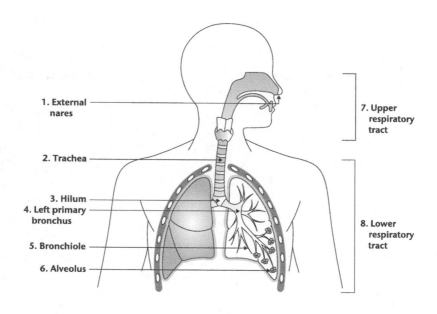

1. External nares
2. Trachea
3. Hilum
4. Left primary bronchus
5. Bronchiole
6. Alveolus
7. Upper respiratory tract
8. Lower respiratory tract

1 *External nares (nostrils):* contain numerous short hairs which filter out dust and large foreign particles. Air travels to the two nasal passages, which are separated by the septum. Interior structures of the nose are specialized for three functions: (1) incoming air is warmed, moistened and filtered; (2) olfactory stimuli are received; and (3) large, hollow, resonating chambers modify speech sounds.

2 *Trachea (windpipe):* the essential passageway for air. Approximately 5 cm long, 2.5 cm diameter, located in front of the oesophagus. Extends from the larynx to T5 vertebral area, where it divides into right and left primary bronchi.

3 *Hilum:* the region where the bronchus, blood vessels, nerves and lymphatics enter the pleural cavities and lungs.

4 *Left primary bronchus:* supplies air to the left lung. The left primary bronchus is longer, narrower and more horizontal than the right primary bronchus. Inhaled foreign objects have a tendency to lodge in the right bronchus rather than the left because of this.

5 *Bronchiole:* consists of terminal bronchioles and the acinus (respiratory bronchioles, alveolar ducts, alveolar sac and individual alveoli).

6 *Alveolus (pl. alveoli):* the site of gas exchange in the lungs. Type I cells form the continuous lining of the alveolar wall and are the main site of gas exchange. Type II cells produce the alveolar fluid (surfactant) which keeps the alveoli moist and helps decrease surface tension to keep alveoli inflated. Together this maintains normal, easy breathing. Alveolar macrophages (dust cells) remove fine dust and other debris that reach the alveolar spaces.

Additional mechanisms are in place throughout the respiratory system to prevent foreign material from entering the lungs, for example, mucous membranes, ciliated epithelium, and the sneeze/coughing reflex.

7 *Upper respiratory tract:* the nose, nasal conchae, oropharynx, laryngopharynx, uvula, epiglottis, vocal cords and associated cartilages (and associated structures).

8 *Lower respiratory tract:* the trachea, bronchi, bronchioles, alveolar ducts and alveoli. The lower tract can be further divided into the conducting portion and the respiratory airways. The conducting elements are the trachea, bronchi and bronchioles (down to the level of the terminal bronchioles). C-shaped cartilage rings reinforce the trachea and prevent it from collapsing. Rings of cartilage are also present in the larger bronchi which also possess smooth muscle and epithelium. As the bronchi branch out and become smaller, they lose cartilage and then smooth muscle – the smallest bronchioles consist of a single layer of epithelial cells. The respiratory portion consists of the respiratory bronchioles, alveolar ducts and the alveolar sac containing many alveoli. Collectively, these are known as an acinus.

TRUE OR FALSE?

9 **Free movement of a gas always proceeds from a region of high pressure to a region of lower pressure.** ✓

This is because a pressure gradient exists. Air moves into the lungs (inspiration) when the pressure inside the lungs is less than the air pressure in the atmosphere. Air moves out of the lungs (expiration) when the pressure inside the lungs is greater than the pressure in the atmosphere. Inspiration involves the contraction of the main inspiratory muscles – the diaphragm and intercostals. Contraction of these muscles increases the volume of the thoracic cavity, reducing the pressure within the lungs. Relaxation of these muscles reduces the volume of the thoracic cavity, therefore increasing lung pressure above atmospheric pressure. Due to this pressure gradient, air moves out of the lungs into the atmosphere.

10 **Pulmonary ventilation is the exchange of gases between the atmosphere and alveoli.** ✓

Pulmonary ventilation (breathing) involves movement of air between the atmosphere and lungs and occurs because of the pressure gradient. The diaphragm and the intercostal muscles allow expansion of the thoracic and abdominal cavities. The diaphragm is a dome-shaped sheet of muscle located below the lungs that separates the thoracic and abdominal cavities. When it contracts, it moves down, the dome flattens and the size of the chest cavity increases, lowering pressure inside the lungs. The intercostal muscles are located between the ribs; when they contract, the ribs move up and outwards. Their action also increases the size of the chest cavity and lowers the pressure inside the lungs. By contracting, the diaphragm and intercostal muscles reduce the internal pressure relative to the atmospheric pressure, allowing air into the lungs. The Hering–Bruer reflex exists to protect the lungs from over-inflation and is triggered by pulmonary stretch receptors, which, when stimulated, send impulses to the inspiratory area of the medulla and the apneustic area of the pons, thus inhibiting further intake of air. During exhalation, the reverse occurs. The diaphragm relaxes and its dome curves up into the chest cavity, while the intercostal muscles relax and bring the ribs down and inward. The reduced size of the chest cavity increases the pressure in the lungs, forcing air out.

11 **The pO_2 in alveolar air is approximately 13.3 kPa.** ✓

The partial pressure of O_2 (pO_2) within alveolar air is less than that of the pO_2 in the atmosphere (approximately 21.3 kPa) because O_2 is continually moving out the alveoli into the pulmonary circulation (which is low in O_2 and high in CO_2 at the alveoli). The pO_2 of venous blood is approximately 5.3 kPa. O_2 therefore moves from the alveoli into the blood. The pCO_2 in the body tissues is approximately 5.8 kPa. Within the systemic capillaries, the pCO_2 is approximately 5.3 kPa and therefore CO_2 enters the bloodstream from the tissues by diffusion (from high concentration to lower concentration). At the same time, O_2 moves from the capillaries (pO_2 = 13.3 kPa) to the tissues (pO_2 = 5.3 kPa).

12 | **The right primary bronchus is more horizontal than the left.**

The right primary bronchus is more vertical, shorter and wider than the left. It supplies the tri-lobed right lung, has incomplete rings of cartilage (like the trachea) and is lined by ciliated epithelium. The function of this epithelium is to sweep the mucus and foreign material towards the larynx.

13 | **The basic respiration rhythm is controlled by parts of the nervous system in the medulla oblongata and pons.**

At rest, about 200 mL of O_2 are used each minute. During strenuous exercise, demand for O_2 can increase 30-fold and nervous mechanisms exist to match respiratory effort with metabolic demand.

14 | **Lung capacity increases with age.**

With advancing years, the airways and tissues of the respiratory tract become less elastic and more rigid, as does the chest wall, causing a decrease in lung capacity, typically 35 per cent by age 70.

15 | **A peak flow meter measures ability to inhale air.**

A peak flow meter is a hand-held device that measures peak expiratory flow rate (PEFR or PEF) by gauging the speed of exhaling (breathing out) air by measuring airflow through the bronchi. This determines the degree of obstruction in the airways. Peak flow readings are high when airways are not obstructed and lower when the airways are constricted. It is often used to monitor respiratory disorders such as asthma or chronic obstructive pulmonary disease (COPD). (See *Paramedics! Test Yourself in Pathophysiology*.)

16 | **The respiratory system is involved in regulating blood pH.**

Along with the kidneys, the respiratory system is very important in maintaining blood pH within the acceptable 7.35–7.45 range. The respiratory system does this by controlling the amount of carbon dioxide (CO_2) in the blood. Carbon dioxide is acidic in the blood, so if it accumulates in the body, the blood may become dangerously acidic. Since the lungs are responsible for removing CO_2 during exhalation, they play a vital role in regulating blood pH. This is why certain respiratory disorders can cause a life-threatening change in blood pH.

17 | **The epiglottis serves to protect the lower respiratory tract from foreign matter.**

The epiglottis is a leaf-shaped elastic cartilage covered by a mucous membrane. It guards the entrance to the glottis opening and in its relaxed position it sits upright. When an individual swallows, the hyoid bone rises and the larynx moves with it. At the same time the epiglottis becomes more horizontal and occludes the glottic opening, thus serving to protect the lower respiratory tract from foreign matter.

18 **The epiglottic vallecula is part of the lower respiratory tract.**

Vallecula is a term used to describe any anatomical crevice or depression. In this case it describes the depression between the base of the tongue and the epiglottis and as such is part of the upper respiratory tract. In paramedic practice this space is of importance as it is the location where placement of a Macintosh laryngoscope blade is placed during endotracheal intubation.

19 **The cricothyroid membrane is constructed of fibrolastic tissue.**

The cricothyroid membrane is made up three ligaments: a central ligament attaching to the thyroid and cricoid cartilages, with a pair of lateral cartilages attaching medially. The purpose of these membranes is to inhibit both the thyroid and cricoid cartilages from excessive movement. Identification of this membrane has an important relevance when carrying out needle cricothyroidotomy.

MULTIPLE CHOICE

Correct answers identified in bold italics.

20 **Hypoxia is a tissue-level deficiency of:**

a) oxygen   b) carbon dioxide   c) carbon monoxide   d) water

Hypoxia means 'low oxygen'. There are a number of types of hypoxia – hypoxic hypoxia (caused by low arterial pO_2), anaemic hypoxia (insufficient haemoglobin to carry O_2), stagnant hypoxia (poor blood flow), histotoxic hypoxia (tissue has adequate oxygen supply but cannot use it, for example, cyanide poisoning).

21 **The pharynx is divided into how many sub-divisions?**

a) 2   *b) 3*   c) 4   d) 5

The pharynx can be divided into the nasopharynx, oropharynx and laryngopharynx. The oropharynx and laryngopharynx are involved in both conduction of air and food. Liquids and foods are routed into the oesophagus by the epiglottis.

22 **The posterior thoracic cage is composed of how many pairs of ribs?**

a) 8   b) 10   *c) 12*   d) 14

The vertebral column and 12 pairs of ribs form the posterior portion of the thoracic cage. The anterior thoracic cage consists of the sternum (manubrium, body and xiphoid process) and ribs. Rib pairs 1 to 7 attach directly to the sternum (true ribs), rib pairs 8 to 12 are false ribs (pairs 8 through 10 attach to the cartilage of the preceding rib). Pairs 11 and 12 are called 'floating ribs' because they do not attach to any part of the anterior thoracic cage.

23 **The precise site of gas exchange within the lungs occurs at:**

a) the nasal mucosa   b) the bronchi   c) the bronchioles
d) the alveoli (sing. alveolus)

The exchange of respiratory gases between the lungs and blood occurs via diffusion across alveolar and capillary walls (respiratory membrane). The alveoli have a number of special adaptations that facilitate their essential role in gas exchange: (1) they are only one cell thick – therefore a very short diffusion distance; (2) they are bathed in surfactant – moisture aids diffusion of gases; (3) they are surrounded by a vast capillary network – provides good blood supply to move gases around.

24 **The wandering phagocytes found in the alveoli are called:**

a) alveolar macrophages   b) pulmonary cells   c) goblet cells
d) chalice cells

Also known as 'dust cells', they remove dust and other debris from the alveoli by phagocytosis. Some particles (for example, asbestos, silica) cannot be engulfed and digested by alveolar macrophages, resulting in irreversible respiratory disease through an accumulation of these materials within lung tissue.

25 **Carbon dioxide is carried from body tissues in what form?**

a) dissolved CO_2 b) oxyhaemoglobin   c) carbaminohaemoglobin
d) bicarbonate ions

Bicarbonate ions are also known as hydrogencarbonate ions (HCO_3^-). These are alkaline and have a vital role in maintaining acid–base homeostasis in the body. They provide resistance to dramatic changes in pH. This is particularly important in the CNS tissues where pH deviation too far beyond the normal range can be life-threatening. The blood value of bicarbonate ions is one of several indicators of acid–base physiology in the body. Bicarbonate also regulates pH in the small intestine. This is released from the pancreas in response to the hormone secretin to neutralize the acidic chyme entering the duodenum from the stomach.

26 Lung volumes exchanged during breathing are measured using:

a) a stethoscope **b) a spirometer** c) a sphygmomanometer
d) a bronchoscope

This apparatus is commonly used to measure the volume of air exchanged during breathing and the rate of ventilation (Figure 8.3). A stethoscope is used to listen to internal sounds of the body, usually the heartbeat and the lungs. A sphygmomanometer (blood pressure cuff) is the instrument used to measure blood pressure. A bronchoscope is used to visualize the airways in the procedure known as bronchoscopy. In clinical practice, the term respiration (ventilation) means one inspiration and one expiration. A healthy adult averages 12 respirations per minute and moves approximately 6 litres of air in and out of the lungs when resting. A low respiratory volume usually indicates pulmonary malfunction.

Figure 8.3 Respiration trace from a spirometer

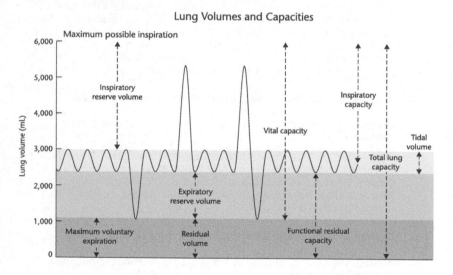

27 The total lung capacity in a healthy adult male is approximately:

a) 500 mL   b) 5000 mL   **c) 6000 mL**   d) 6000 L

Lung capacities are combinations of specific lung volumes. Inspiratory capacity, the total inspiratory ability of the lungs, is the sum of the tidal volume plus inspiratory reserve volume (3600 mL). Functional residual capacity is the sum of residual volume plus expiratory reserve volume (2400 mL). Vital capacity is the sum of inspiratory reserve volume, tidal volume and expiratory reserve volume (4800 mL). The total lung capacity is the sum of all volumes. (Note: take care when working with units, notice the subtle difference between options *c* and *d* – yet there is a thousand-fold difference in these two volumes!)

FILL IN THE BLANKS

28 | **Blood travels from the pulmonary arteries to the *lungs*.**

Blood in the pulmonary artery is considered to be deoxygenated when leaving the heart because it is low in oxygen (pO_2= 5.3 kPa) in comparison to blood in the highly oxygenated pulmonary vein (pO_2= 13.3 kPa). In the pulmonary artery, blood travels directly to the capillary bed of the lungs where it becomes oxygenated. It travels directly back to the heart in the pulmonary vein, entering the left atrium of the heart.

29 | **Most of the oxygen transported in blood is bound to *haemoglobin*.**

Approximately 98.5 per cent oxygen within blood is carried as oxyhaemoglobin. The remainder is dissolved in plasma. CO_2 is carried in the blood in three forms: (1) bicarbonate ions (approximately 70 per cent); (2) carbaminohaemoglobin (approximately 23 per cent); and (3) dissolved CO_2 (approximately 7 per cent).

30 | ***Bronchoscopy* is the visual examination of the bronchus through a bronchoscope.**

Bronchoscopy is a technique of visualizing inside the airways for diagnostic and therapeutic purposes. The bronchoscope is inserted into the airways, usually through the nose or mouth, or occasionally through a tracheostomy. It allows examination of a patient's airways for abnormalities such as foreign bodies, bleeding, tumours, or inflammation. Specimens may be taken from inside the lungs: biopsy, fluid or endobronchial brushing.

31 | **The vocal cords are contained within the *larynx*.**

The larynx (sometimes called the voice box) contains the vocal cords and connects the laryngopharynx with the trachea. Muscle and cartilage form the walls of the larynx. Sound is generated in the larynx and it is where pitch and volume are manipulated. The strength of expiration from the lungs also contributes to the loudness of the sound produced. Pitch is influenced by the length and thickness of the vocal cords and the tightening and relaxation of the muscles surrounding them.

32 | **The production of *surfactant* by type II alveolar cells reduces surface tension within the lungs.**

Surfactant is a detergent-like substance which reduces the tendency of alveoli to collapse inwards due to the forces of surface tension. Surface tension arises at air–liquid junctions.

33 **The lungs are contained within the *thoracic* cavity.**

The lungs are separated from each other by the heart and other structures in the mediastinum. The mediastinum separates the thoracic cavity (thorax, chest cavity) into two distinct chambers which allows one lung to remain expanded should the other collapse. Two layers of pleural (or serous) membranes enclose and protect each lung.

34 **The exchange of gases between alveoli and their surrounding capillaries is called *external* respiration.**

External respiration occurs between alveoli (where gas exchange occurs in the lungs) and the external atmosphere surrounding the body. Internal respiration occurs between the circulating blood and the body's tissues.

35 **Healthy adults take approximately *12* breaths per minute.**

This rate changes with exercise and with age. Newborns take an average of 45 breaths per minute although this value decreases during childhood reaching approximately 12–15 breaths per minute in the normal resting adult. In exercising adults this value increases to approximately 35–45 breaths per minute. Respiration rates increase with fever, illness or other medical conditions.

During normal quiet breathing, we exhale passively as the lungs recoil and the muscles relax. For rapid and deep breathing, however, the expiratory centre in the brain becomes active and sends impulses to the muscles to force exhalations. The normal breathing rate changes to match the body's needs. The rate and depth of breathing can be consciously controlled. The voluntary control of breathing allows the 'holding of breath', for example, avoiding breathing in water or harmful chemicals for brief periods of time. It is, however, impossible to consciously stop breathing for a prolonged period because an accumulation of carbon dioxide in the blood activates the brain's breathing centre (despite voluntary control) to take a gasp of breath – therefore people cannot die by simply holding their breath.

36 ***Compliance* is the ease with which the lungs and thoracic wall expand.**

Pulmonary compliance (or lung compliance) is the ability of the lungs to stretch during a change in volume relative to a change in pressure. Compliance will be affected by certain pulmonary diseases. Fibrosis of the lungs makes the lungs stiffer, hence, decreasing lung compliance. In emphysema the lungs becoming loose and floppy, so only a small pressure difference is necessary to maintain a large volume resulting in an increase in lung compliance.

37 **The respiratory centre consists of a medullary rhythmicity area, *pneumotaxic* area and an apneustic area.**

The respiratory centre in the brain controls the mechanical action of the lungs. The size of the thorax is affected by the action of the respiratory muscles which contract and relax due to nervous stimulation from the brain. The area which sends nerve impulses to respiratory muscles is located in the medulla oblongata and pons of the brain stem. It consists of a widely dispersed group of neurones that are functionally divided into three areas: (1) the medullary rhythmicity centre in the medulla oblongata; (2) the pneumotaxic area in the pons; and (3) the apneustic area, also in the pons.

9 The digestive system

INTRODUCTION

There are seven components of the diet: carbohydrates, lipids, proteins, vitamins, minerals, fibre and water. The digestive system prepares food nutrients for use by the body's cells since the nutrients are unable to directly enter cells because they cannot pass through the intestinal walls to the bloodstream.

The digestive system includes all the organs and glands involved in eating and digesting. Its function is to break down food and absorb nutrients. There are two basic divisions to the digestive system: the gastrointestinal (GI) tract (or alimentary canal) and the accessory digestive organs.

The digestive system carries out six basic functions: ingestion, secretion, propulsion, digestion, absorption and defecation.

The GI system is controlled by a range of nervous and hormonal messages and relies on an adequate supply of blood from the cardiovascular system. The GI system is intimately related to these other organ systems, therefore a thorough understanding is essential for paramedics when diagnosing and treating disorders of the GI system.

Useful resources

Paramedics! Test Yourself in Pathophysiology
Chapter 9

Anatomy and Physiology (8th edition)
Chapter 21

Anatomy and physiology of the digestive system:
http://digestive.niddk.nih.gov/ddiseases/pubs/yrdd/

LABELLING EXERCISE

1–9 Identify the regions of the human digestive system in Figure 9.1, using the options provided on page 142.

Figure 9.1 The human digestive system
N.B. Figure not to scale

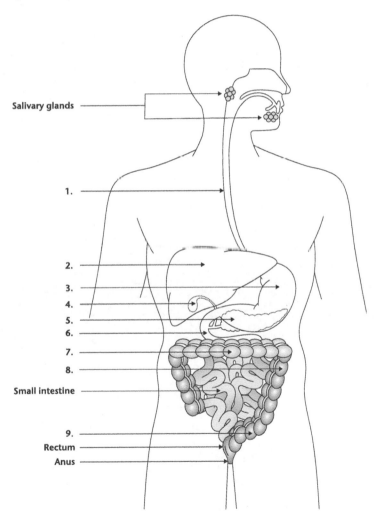

sigmoid colon

oesophagus

gall bladder

pancreas

stomach

duodenum

transverse colon

descending colon

liver

TRUE OR FALSE?

Are the following statements true or false?

10 The oesophagus has both a digestive and respiratory function.

11 The pyloric sphincter prevents the stomach contents from flowing back into the oesophagus.

12 Hydrochloric acid (HCl) is produced by chief cells within the small intestine.

13 Gastric emptying is regulated by neuronal and hormonal reflexes.

14 Ageing causes physiological changes in the GI tract.

15 The retroperitoneal space contains no digestive organs.

MULTIPLE CHOICE

Identify one correct answer for each of the following.

16 The primary phase in the regulation of gastric secretion and motility is called:

a) the swallowing phase
b) the gastric phase
c) the cephalic phase
d) the intestinal phase

17 The enzyme pepsin digests which dietary component?

a) protein
b) fat
c) carbohydrate
d) vitamins

18 Intrinsic factor is produced by which cells within the stomach?

a) mucous cells
b) chief cells
c) enteroendocrine cells
d) parietal cells

19 In the body, the liver is described as:

a) the heaviest and largest organ
b) the heaviest and second largest organ
c) the largest and second heaviest organ
d) the second largest and second heaviest organ

20 The ileum is the longest section of:

   a) the stomach

   b) the liver

   c) the small intestine

   d) the large intestine

FILL IN THE BLANKS

Fill in the blanks in each statement using the options in the box below.
Not all of them are required, so choose carefully!

microvilli	alkaline
villi	sigmoid colon
smooth muscle	urine
ascending colon	water
blood coagulation	faeces
appendix	salts
descending colon	neutral
plicae	transverse colon
rectal	

21 The _____ are circular folds found in the small intestine.

22 The _____ and_____ project into the lumen of the small intestine, increasing surface area.

23 The large intestine can be divided into four distinct colonic regions: the _____ _____, _____ _____, _____ _____, and _____ _____.

24 The main functions of the large intestine are to absorb _____ and _____, and to excrete _____.

25 Bacteria within the large intestine assist in the synthesis of vitamin K which is necessary for _____ _____.

26 The mucosa of the large intestine produce _____ secretions.

ANSWERS

LABELLING EXERCISE

Figure 9.2 The human digestive system

N.B. Figure not to scale

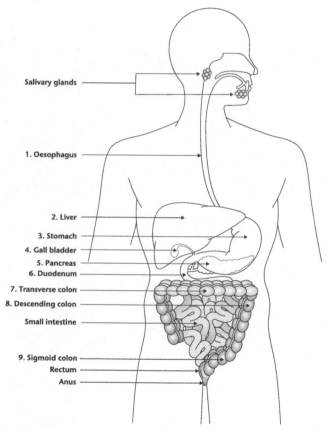

1   *Oesophagus:* a muscular (collapsible) tube lying behind the trachea. Around 25 cm long, it begins at the lower end of the laryngopharynx, passes through the mediastinum, pierces the diaphragm through the hiatus and terminates at the entrance to the stomach. Peristalsis propels solids and liquids down the oesophagus.

2 **Liver:** the heaviest of the accessory digestive organs anchored to the lesser curvature of the stomach by the lesser omentum. It is involved in several important functions: nutrient metabolism, detoxification processes, synthesis of plasma proteins, non-essential amino acids and vitamin A. Also stores essential nutrients (vitamins K, D, B12, and iron), produces urea (from ammonia in body fluids), regulates blood glucose and secretes bile.

3 **Stomach:** a collapsible, J-shaped, structure located in the upper abdominal cavity. It has four regions: (1) its upper region, the cardia – is attached to the lower end of the oesophagus; (2) the fundus – located above and to the left of the oesophageal opening; (3) the body is the middle region of the stomach; and (4) the pylorus is the lower region of the stomach, located near its junction with the duodenum. The stomach serves as a temporary storage of food where digestion commences. Food in the stomach is broken down into chyme (semi-fluid substance) before being moved towards the small intestine. When empty, the mucous membrane lies in longitudinal folds, called rugae, which stretch out smooth when the stomach is full.

4 **Gall bladder:** a pear-shaped accessory organ of the GI tract. It has a capacity of approximately 30 mL. It is located behind the right lobe of the liver and is attached to the liver via the cystic duct which connects with the common hepatic duct. The gall bladder stores and concentrates the bile produced by the liver and releases it into the duodenum via the bile duct which merges with the pancreatic duct as they both connect with the duodenum.

5 **Pancreas:** another accessory organ of the GI tract, located behind the stomach. As a gland, the pancreas has two different types of secretion: (1) internal (or endocrine) secretion – insulin is secreted from the pancreatic islets directly into the blood. This hormone, secreted by the beta cells, is produced when circulating glucose levels rise above the homeostatic threshold by increasing glucose transport and utilisation. Glucagon, however, is released when blood glucose levels fall, by the alpha cells of the pancreas. These cells mobilise energy reserves by causing skeletal muscle cells and the liver to break down glycogen into glucose, also causing an increase in the breakdown of fats to fatty acids and increased synthesis and release of glucose by the liver; (2) external (or exocrine) secretion – digestive enzymes are secreted via the pancreatic duct into the duodenum where they mix with food (chyme) as it leaves the stomach.

6 **Duodenum:** the first and shortest segment of the small intestine at approximately 25 cm long. It starts at the pyloric sphincter of the stomach and ends merging with the jejunum.

7 **Transverse colon:** situated above the small intestine passing horizontally across the abdomen.

8 *Descending colon:* starts near the spleen and extends down the left side of the abdomen into the pelvic cavity.

9 *Sigmoid colon:* descends through the pelvic cavity where it becomes the rectum.

TRUE OR FALSE?

10 **The oesophagus has both a digestive and respiratory function.**

The oesophagus is only concerned with the transport of solids and liquids into the stomach through the process of peristalsis. Swallowing a bolus of food occurs in three stages: (1) the voluntary stage in which the bolus is moved into the oropharynx; (2) the pharyngeal stage which involves the involuntary movement of the bolus through the pharynx into the oesophagus; and (3) the oesophageal stage which causes the involuntary movement of the bolus through the oesophagus into the stomach.

11 **The pyloric sphincter prevents the stomach contents from flowing back into the oesophagus.**

The pyloric sphincter (or pylorus) is a ring of smooth muscle located at the opening between the stomach and the duodenum. The cardiac sphincter is located at the entrance of the stomach at the base of the oesophagus. It prevents the back-flow of stomach contents into the oesophagus. If this fails to occur, heartburn may arise (although this is not related to a cardiac problem). The cardiac sphincter relaxes during swallowing to allow food to enter the stomach. If this fails to occur, a condition called achalasia may occur. Food can become trapped in the oesophagus resulting in severe chest pain which is often confused with that originating from the heart.

12 **Hydrochloric acid (HCl) is produced by chief cells within the small intestine.**

The parietal cells produce HCl and intrinsic factor. HCl protects the small intestine against bacteria, denatures proteins, stimulates the secretion of hormones that promote the flow of bile and pancreatic juice (secretin and cholecystokinin (CCK)), and helps to activate pepsinogen into pepsin.

13 **Gastric emptying is regulated by neuronal and hormonal reflexes.**

Initiation of gastric emptying occurs through stimuli that include distension of the stomach, the presence of partially digested proteins, alcohol and caffeine. These stimuli increase secretion of gastrin and promote vagus nerve impulses causing contraction of the cardiac sphincter and relaxation of the pyloric sphincter. The result of these combined processes is gastric emptying.

14 **Ageing causes physiological changes in the GI tract.**

These changes are considered less debilitating than in other body systems. Changes include: (1) reduced secretions within the GI tract causing changes in digestion and absorption; and (2) reduced motility and muscle tone which can lead to loss of appetite and/or constipation in the older adult.

15 **The retroperitoneal space contains no digestive organs**

Structures contained within the retroperitoneum are described as either primary or secondary organs. The primary organs are the ureters, urinary bladder, aorta, inferior vena cava, part of the oesophagus and a portion of the rectum. The secondary organs include the pancreas with the exception of its tail, the medial and distal portion of the duodenum and the ascending and descending portions of the colon. The factor that determines whether an organ is defined as primary or secondary is that secondary organs have migrated to the retroperitoneal space during embryogenesis, having originated in the peritoneal cavity.

 MULTIPLE CHOICE

Correct answers identified in bold italics.

16 **The primary phase in the regulation of gastric secretion and motility is called:**

a) the swallowing phase　　b) the gastric phase　　*c) the cephalic phase*　　d) the intestinal phase

The cephalic phase begins with reflexes stimulated by sensory receptors in the head in response to smell, taste or thoughts about food. Parasympathetic stimulation of parietal, chief and mucous cells increases secretions from all gastric glands. Stimulation by the parasympathetic system also stimulates smooth muscle within the stomach, resulting in increased gastric motility.

17 **The enzyme pepsin digests which dietary component?**

a) protein　　b) fat　　c) carbohydrate　　d) vitamins

The inactive precursor, pepsinogen, prevents pepsin from digesting the chief cells in the stomach which produce it. Pepsinogen is only converted to active pepsin when it comes into contact with active pepsin molecules or HCl.

18 **Intrinsic factor is produced by which cells within the stomach?**

a) mucous cells   b) chief cells   c) enteroendocrine cells
d) parietal cells

Parietal (or oxyntic) cells also produce HCl. Intrinsic factor is an essential enzyme for absorption of vitamin B_{12} (known as extrinsic factor), which is required for red blood cell formation. A lack of intrinsic factor can cause pernicious anaemia, an autoimmune condition caused by destruction of the parietal cells.

19 **In the body, the liver is described as:**

a) the heaviest and largest organ   *b) the heaviest and second largest organ*   c) the largest and second heaviest organ   d) the second largest and second heaviest organ

The liver is the heaviest organ (approximately 1.5 kg) and is second only to the skin in size. (For functions, see Answer 2.)

20 **The ileum is the longest section of:**

a) the stomach   b) the liver   *c) the small intestine*   d) the large intestine

The small intestine is about 5.5 m long and is the longest organ of the GI tract. It has three major divisions: (1) the duodenum – the shortest region; (2) a middle portion, the jejunum which is 1–2 m long; and (3) the ileum which is the longest portion at approximately 2–4 m long.

FILL IN THE BLANKS

21 **The _plicae_ are circular folds found in the small intestine.**

These help mix chyme and move it along the small intestine. Unlike the rugae in the stomach, the plicae do not smooth out when the small intestine is full. They increase the surface area for absorption to occur and are located all along the small intestine from the proximal end of the duodenum and terminating near the middle portion of the ileum.

22 **The _villi_ and _microvilli_ project into the lumen of the small intestine, increasing surface area.**

The villi are tiny, finger-like projections of the small intestine lining that increase the surface area for absorption and have a network of capillaries and lacteals that aid absorption of nutrients from the chyme. The microvilli are found on the surface of the villi, further increasing the surface area for nutrient absorption. Together villi and microvilli are sometimes called the 'brush border'.

23 **The large intestine can be divided into four distinct colonic regions: the *ascending colon*, *transverse colon*, *descending colon*, and *sigmoid colon*.**

These structures are part of the large intestine (colon) which has a number of functions: it completes the absorption process (initiated in the small intestine), the manufacture of certain vitamins, and the formation and expulsion of faeces from the body. The large intestine produces no hormones or digestive enzymes. It absorbs most of the water in the colon leaving approximately 100 mL remaining. It also absorbs large amounts of sodium and chloride ions.

24 **The main functions of the large intestine are to absorb *water* and *salts*, and to excrete *faeces*.**

Once it enters the large intestine, chyme is not digested any further. Passage through the large intestine is the final part of digestion and takes approximately 10–12 hours to complete in a healthy individual. If passage through the colon is very slow, excess water can be absorbed which causes constipation. Alternatively, if passage is too fast, insufficient water is absorbed causing diarrhoea. If diarrhoea is prolonged, a patient may become dehydrated, losing vital electrolytes as well as water.

25 **Bacteria within the large intestine assist in the synthesis of vitamin K which is necessary for *blood coagulation*.**

Vitamin K is a fat-soluble vitamin produced by the intestinal bacteria. A deficiency results in increased clotting time which may result in excessive bleeding.

26 **The mucosa of the large intestine produce *alkaline* secretions.**

These thick, alkaline secretions are produced by the goblet cells of the intestinal walls. They lubricate and neutralize the faeces, easing their passage through the intestine and reducing erosion of the intestinal lining.

153

10 The urinary system

INTRODUCTION

Structurally, the urinary system is comprised of two kidneys, two ureters, a urinary bladder and a urethra. The kidneys are primarily involved in the management and removal of waste but they have a number of additional functions. These include maintaining homeostasis by regulating the composition, volume and pressure of blood; regulating blood pH and contributing to metabolism.

The kidneys filter blood and return selected amounts of water and solutes to the blood. The remaining water and solutes constitute urine.

Urine is excreted from each kidney through its associated ureter and stored in the urinary bladder until it is removed from the body via the urethra during micturition (urination). In the male, the urethra also has a reproductive function acting as a channel for the discharge of seminal fluid. Maintaining fluid balance in the body is an essential part of homeostasis, therefore understanding how the urinary system is involved in this is an essential part of paramedic practice.

Useful resources

Paramedics! Test Yourself in Pathophysiology

Chapter 10

Anatomy and Physiology (8th edition)

Chapter 23

Interactive tutorial on the urinary system:

http://www.getbodysmart.com/ap/urinarysystem/menu/menu.html

LABELLING EXERCISE

1–10 Identify the regions and structures of the human kidney in Figure 10.1, using the options provided in the box below.

renal pelvis   renal artery

renal calyx   renal medulla

ureter   renal column

renal cortex   renal capsule

renal pyramid   renal vein

Figure 10.1 The kidney

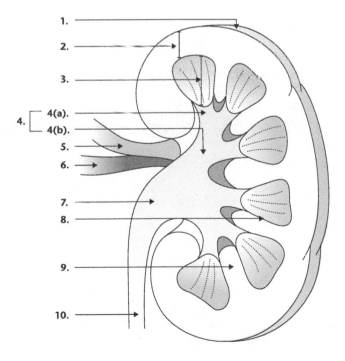

TRUE OR FALSE?

Are the following statements true or false?

11 The normal pH of urine lies in the range 7.35–7.45.

12 The kidneys are surrounded by two distinct layers of tissue.

13 A frontal section through a kidney reveals two distinct regions.

14 The functional units of the kidney are called the renal pyramids.

15 Renal corpuscles are found in the renal medulla.

16 The filtering unit of a nephron is called the proximal convoluted tubule.

17 Blood proteins are readily filtered through the glomerular filtration apparatus.

18 The afferent arterioles supplying renal corpuscles have a larger diameter than the efferent arterioles.

19 Tubular reabsorption is a non-selective process.

20 Antidiuretic hormone and aldosterone regulate water and solute reabsorption in the descending loop of Henlé.

21 Adrenaline, produced by the kidneys, is involved in erythrocyte production.

22 The renin-angiotensin system regulates urine output.

MULTIPLE CHOICE

Identify one correct answer for each of the following.

23 Approximately what volume of filtrate enters the glomerular capsule space each day?

a) 80 L

b) 130 L

c) 180 L

d) 230L

24 The blood filtering capacity of the renal corpuscles is enhanced by:

a) the thin porous endothelial-capsular membrane

b) a large capillary surface area

c) high capillary pressure

d) all of the above

25 Through which arteriole does blood exit the glomerular capsule?

a) afferent

b) efferent

c) renal

d) interlobular

26 The group of modified cells lying adjacent to the afferent and efferent arterioles is called:

a) the renal corpuscle

b) the juxtaglomerular apparatus

c) the peritubular network

d) the vasa recta

27 Which of the following acts as the filtration apparatus of the kidneys?

a) the descending loop of Henlé

b) the glomerular capsule

c) the collecting duct

d) the renal pelvis

28 Which of the following structures is not part of the nephron?

a) the calyx

b) the distal convoluted tubule

c) the ascending loop of Henlé

d) the collecting duct

29 What does the urinary bladder do?

a) secretes urine

b) stores urine

c) concentrates urine

d) dilutes urine

30 Which of the following substances does not normally pass through the glomerular capsule?

a) albumin

b) glucose

c) urea

d) sodium ions

31 The process of voiding urine is termed:

a) menstruation

b) filtration

c) micturition

d) fibrillation

32 Anuria is best defined as:

a) blood in the urine

b) failure to release urine

c) excessive urine production

d) excessive urine production at night

33 Creatinine is a metabolic waste product excreted in urine and derived from:

a) liver

b) muscle

c) bone

d) skin

34 Normal urine:

a) is usually alkaline

b) contains approximately 75 per cent water

c) has a specific gravity around 1.001–1.035

d) has a high protein content

MATCH THE TERMS

Match each term with the correct description.

A. glomerular filtration   **F.** afferent arteriole

B. aldosterone   **G.** renal cortex

C. renal medulla   **H.** tubular secretion

D. antidiuretic hormone   **I.** juxtaglomerular apparatus

E. nephron   **J.** tubular reabsorption

| **35** | Adrenal hormone that helps regulate fluid volume _____ |

| **36** | Functional unit of the kidney _____ |

| **37** | Primary phase in urine formation _____ |

| **38** | The inner region of the kidney _____ |

| **39** | Movement of substances from the peritubular capillary to the glomerular filtrate _____ |

| **40** | Outer region of the kidney _____ |

| **41** | Vessel that conveys blood to the glomerulus _____ |

| **42** | Anterior pituitary hormone that enhances water conservation _____ |

| **43** | Movement of substances from the glomerular filtrate into the peritubular capillary _____ |

| **44** | Mass of cells located in the walls of glomerular arterioles that help regulate renal blood flow _____ |

ANSWERS

LABELLING EXERCISE

Figure 10.2 The kidney

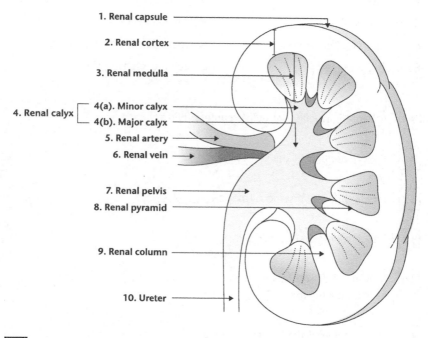

1. Renal capsule
2. Renal cortex
3. Renal medulla
4. Renal calyx
   - 4(a). Minor calyx
   - 4(b). Major calyx
5. Renal artery
6. Renal vein
7. Renal pelvis
8. Renal pyramid
9. Renal column
10. Ureter

1 *Renal capsule:* deep layer of tissue surrounding each kidney. It protects and maintains the kidney shape.

2 *Renal cortex:* superficial outer area of the kidney. Contains the initial filtering portion of the nephron – the renal corpuscle.

3 *Renal medulla:* deep inner portion of the kidney, containing the renal tubule portion of the nephron. The nephrons span the renal cortex and renal medulla.

4 *Renal calyx:* structural component of the kidney into which urine ultimately drains. Each kidney has 8–18 minor calyces and 2–3 major calyces. From the major calyx, urine drains into the renal pelvis and then out through the ureter to the urinary bladder.

5 | **Renal artery:** vessel supplying the kidney with blood. Although the kidneys make up less than 1 per cent of the total body mass, they receive 20–25 per cent of the resting cardiac output through the left and right renal arteries.

Each nephron has one afferent arteriole which divides into the glomerular capillary network which later unite, forming an efferent arteriole which drains blood from the glomerulus. This vessel system is unique as blood usually flows from capillaries into venules and not other arterioles.

6 | **Renal vein:** vessel through which blood leaves the kidney and which exits at the hilus. Blood flow through the kidney is detailed in Figure 10.3.

Figure 10.3 Renal blood flow

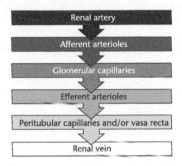

7 | **Renal pelvis:** cavity in the centre of the kidney formed by part of the ureter lying within the kidney and into which the major calyces open.

8 | **Renal pyramid:** a triangular, cone-shaped structure in the renal medulla containing the straight segments of renal tubules and the vasa recta (capillaries that branch from the efferent arterioles leaving each glomerulus). The base of each pyramid faces the renal cortex and its apex points towards the centre of the kidney.

9 | **Renal column:** portions of the renal cortex that extend between the renal pyramids.

10 | **Ureter:** a tube that connects each kidney with the urinary bladder. There are two ureters: one for each kidney. Each ureter is approximately 25 cm long and connects the renal pelvis of one kidney to the bladder. A valve formed by the compression of the bladder as it fills prevents back-flow of urine into the ureters.

TRUE OR FALSE?

11 **The normal pH of urine lies in the range 7.35–7.45.**

The pH of urine ranges between 4.6 and 8.0, with an average of 6.0 although this varies considerably with diet. High protein diets increase acidity while vegetarian diets typically increase alkalinity.

12 **The kidneys are surrounded by two distinct layers of tissue.**

Three layers of tissue surround each kidney. The deep layer is the renal capsule comprising of a dense fibrous membrane. The intermediate layer is formed from adipose tissue and the outer layer is composed of dense irregular connective tissue. Together these tissue layers protect the kidneys from trauma and maintain their position within the abdominal cavity.

13 **A frontal section through a kidney reveals two distinct regions.**

A frontal section through a kidney reveals two distinct regions: a superficial reddish area – the renal cortex, and a deep reddish-brown region – the renal medulla.

14 **The functional units of the kidney are called the renal pyramids.**

The renal pyramids are triangular structures found in the renal medulla containing the straight segments of renal tubules and the vasa recta. The functional units of the kidneys are the nephrons which produce urine.

15 **Renal corpuscles are found in the renal medulla.**

The renal medulla contains the cone-shaped renal pyramids. Renal corpuscles lie in the renal cortex. Each corpuscle has two components: the glomerulus (capillary network) and the glomerular (Bowman's) capsule containing the glomerulus.

As blood flows through the glomerular capillaries, water and most solutes filter from blood plasma into the capsule space. Large plasma proteins and the formed elements of blood do not normally pass through this filter.

16 **The filtering unit of a nephron is called the proximal convoluted tubule.**

The filtering unit of the nephron is the renal corpuscle. The proximal convoluted tubule consists of a coiled tubule connected to the glomerular capsule and the loop of Henlé. In the proximal convoluted tubule, the filtrate is modified: sodium and other solutes are reabsorbed by active transport; and water lost through filtration is reabsorbed by osmosis.

17 **Blood proteins are readily filtered through the glomerular filtration apparatus.**

Blood proteins and the formed elements of blood are not normally filtered through the glomerular filtration apparatus. However, disease processes may directly or indirectly affect the structure, therefore allowing such molecules to pass through.

18 **The afferent arterioles supplying renal corpuscles have a larger diameter than the efferent arterioles.**

The afferent arterioles have a much larger diameter than the efferent arterioles. This increases blood pressure in the glomerular capillaries more than in capillaries elsewhere in the body. The increased pressure enhances the blood-filtering capacity.

19 **Tubular reabsorption is a non-selective process.**

Tubular reabsorption moves substances from the tubular fluid back into the blood within the peritubular capillary. The process involves the selective return of water and solutes such as glucose, amino acids and essential electrolytes such as sodium, potassium, calcium, chloride, bicarbonate and phosphate. Small proteins and peptides that pass through the glomerular filter are also reabsorbed. Waste products and excess substances are not returned to the blood and exit the body.

Tubular secretion, the opposite of reabsorption, moves unwanted substances from blood within the peritubular capillary into the renal tubule for secretion in urine. Secreted substances include hydrogen, potassium and ammonium ions, creatinine and certain drugs. Tubular secretion has two main effects: (1) it helps maintain blood pH; and (2) it rapidly eliminates certain substances from the body.

20 **Antidiuretic hormone and aldosterone regulate water and solute reabsorption in the descending loop of Henlé.**

Antidiuretic hormone (ADH) regulates water reabsorption from the distal convoluted tubule (DCT) into surrounding capillaries while aldosterone controls solute reabsorption. Approximately 90 per cent of the water lost during filtration is reabsorbed by osmosis, together with reabsorption of solutes such as glucose and electrolytes (ions). The remaining 10 per cent is regulated exclusively by ADH.

21 **Adrenaline, produced by the kidneys, is involved in erythrocyte production.**

The hormone involved in the production of erythrocytes is erythropoietin. It is released by the kidneys (and to a lesser extent by the liver) in response to prolonged oxygen deficiency. It stimulates increased production of red cell precursors by the red bone marrow.

Adrenaline (epinephrine) is a hormone secreted by the adrenal medulla that produces actions similar to those that result from sympathetic nervous stimulation.

22 **The renin-angiotensin system regulates urine output.**

The renin-angiotensin system (RAS) also known as the renin-angiotensin-aldosterone system, is a hormone system that regulates blood pressure and water (fluid) balance. When blood volume is low, the kidneys secrete the enzyme renin (do not confuse with *rennin* – a milk-related enzyme) which stimulates production of angiotensin. Angiotensin triggers constriction of blood vessels, increasing blood pressure. It also stimulates aldosterone secretion from the adrenal cortex. Aldosterone causes the kidney tubules to increase the reabsorption of sodium and water back into the blood. This increases the volume of fluid in the body, which also increases blood pressure. If the RAS is hyperactive, blood pressure will increase; however, many drugs are available that disrupt different phases in the RAS and lower blood pressure. This is how 'ACE inhibitor' drugs control high blood pressure (hypertension).

 MULTIPLE CHOICE

Correct answers identified in bold italics.

23 **Approximately what volume of filtrate enters the glomerular capsule space each day?**

a) 80 L   b) 130 L   *c) 180 L*   d) 230 L

Glomerular filtrate amounts to about 180 L of fluid per day. This greatly exceeds the entire blood plasma volume. However, most of this filtrate returns to the blood with the exception of around 1–2 L, which is excreted as urine.

24 **The blood filtering capacity of the renal corpuscles is enhanced by:**

a) the thin porous endothelial-capsular membrane   b) a large capillary surface area   c) high capillary pressure   *d) all of the above*

Glomerular capillaries are the most permeable in the body, being about 50 times more porous than other types of capillary. The presence of relatively large pores in the glomerular capillaries prevents the filtration of blood cells but allows all components of plasma through. Large proteins are prevented from filtering into the glomerular fluid by the basement membrane.

Within the glomerulus, the capillary network is extensive, providing a large surface area for filtration. This is complemented by the relatively high glomerular capillary blood pressure.

25 **Through which arteriole does blood exit the glomerular capsule?**

a) afferent   **b) efferent**   c) renal   d) interlobular

Blood leaves the glomerular capsule via the efferent arteriole. The efferent arteriole is smaller in diameter than the afferent arteriole, thus increasing glomerular filtration pressure and hence blood pressure.

26 **The group of modified cells lying adjacent to the afferent and efferent arterioles is called:**

a) the renal corpuscle   **b) the juxtaglomerular apparatus**   c) the peritubular network   d) the vasa recta

The juxtaglomerular apparatus consists of a cluster of cells in the DCT and juxtaglomerular cells (modified cells of the afferent and sometimes efferent arteriole) which together stimulate secretion of the enzyme renin when blood pressure starts to fall.

27 **Which of the following acts as the filtration apparatus of the kidneys?**

a) the descending loop of Henlé   **b) the glomerular capsule**   c) the collecting duct   d) the renal pelvis

The glomerular capsule (Bowman's capsule) is a sac-like structure which surrounds the glomerulus. The capsule expands at the proximal end of the renal tubule and receives filtrate from the glomerular capillary network.

The descending loop of Henlé is the first part of the loop. It is highly permeable to water and relatively impermeable to solutes such as sodium and chloride ions. It is moderately permeable to urea. This allows a strong osmotic gradient to form along the length of the structure, allowing it to concentrate urine.

Collecting ducts allow the contents of several nephrons to drain into the large ducts of renal pyramids which then drain into the minor calyces.

28 **Which of the following structures is not a part of the nephron?**

a) the calyx   b) the distal convoluted tubule   c) the ascending loop of Henlé   d) the collecting duct

The calyx (pl. calyces) surrounds part of the renal pyramids. Urine formed in the kidney passes through a papilla into the minor calyx, then to the major calyx before passing through the renal pelvis into the ureter.

Peristalsis of the smooth muscle in the walls of the calyces propels urine through the renal pelvis and ureters to the bladder.

The ascending and descending loop of Henlé and the proximal convoluted tubule are all essential structures of the nephron.

29 **What does the urinary bladder do?**

a) secretes urine   **b) stores urine**   c) concentrates urine   d) dilutes urine

The urinary bladder is a muscular organ situated in the pelvic cavity behind the pubic symphysis. It stores urine but has no involvement in its composition.

30 **Which one of the following substances does not normally pass through the glomerular capsule?**

a) albumin   b) glucose   c) urea   d) sodium ions

Large proteins such as albumin do not normally pass into the tubular fluid. Small molecules such as glucose, urea and electrolytes readily pass through the membrane into the glomerular filtrate.

31 **The process of voiding urine is termed:**

a) menstruation   b) filtration   **c) micturition**   d) fibrillation

Micturition (urination) is the expulsion of urine from the urinary bladder. When approximately 200–300 mL of urine has collected in the bladder, stretch receptors in its wall initiate the micturition reflex. This stimulates contraction of the bladder muscle tissue and involuntary relaxation of the internal sphincter. Simultaneously, the cerebral cortex raises awareness of the need to urinate. If socially acceptable, the external sphincter is voluntarily relaxed allowing micturition. If socially unacceptable, the external sphincter remains closed. In this situation, the stimuli cease for a short time but soon return. This recurs until micturition takes place. In babies, the conscious element of micturition is absent and the process is controlled only by the reflex action. With conditioning, the voluntary and involuntary processes become combined.

Menstruation relates to the shedding of blood and other tissue from the uterus at the end of the female reproductive cycle. Fibrillation describes the abnormal spontaneous contraction of muscle cells.

32 **Anuria is best defined as:**

a) blood in the urine   **b) failure to release urine**   c) excessive urine production   d) excessive urine production at night

Anuria describes the failure to produce or release urine. It may be due to kidney failure or an obstruction in the urinary pathway. Blood in the urine is haematuria. Excessive urine production is called polyuria. Excessive urination at night is nocturia.

33 **Creatinine is a metabolic waste product excreted in urine and derived from:**

a) liver　***b) muscle***　c) bone　d) skin

Creatinine is a break-down product of creatine phosphate in muscle and is usually produced at a constant rate by the body (depending on muscle mass). Creatinine is chiefly filtered from the blood by the kidneys, although a small amount is actively secreted by the kidneys into the urine. There is little or no tubular reabsorption of creatinine. If the filtering of the kidney is deficient, blood creatinine levels rise. Therefore, creatinine levels in blood and urine may be used to calculate creatinine clearance, which reflects the glomerular filtration rate (GFR). This is clinically important because it is a measurement of renal function.

34 **Normal urine:**

a) is usually alkaline　　b) contains approximately 75 per cent water
c) has a specific gravity around 1.001–1.035　d) has a high protein content

Specific gravity measures the kidneys' ability to concentrate or dilute urine. It measures solute concentration since urine is a solution of minerals, salts and compounds dissolved in water. More concentrated urine will have a higher specific gravity. Healthy adult kidneys can control concentration/dilution of urine but for infants, the specific gravity range is lower because immature kidneys are less effective at concentrating urine. Urine produced from a normal healthy individual has a pH that covers a wide range (pH 4.5–8.0), it should contain no protein and water content should be in excess of 95 per cent.

MATCH THE TERMS

35 Adrenal hormone that helps regulate fluid volume

B. aldosterone

36 Functional unit of the kidney

E. nephron

37 Primary phase in urine formation

A. glomerular filtration

38 The inner region of the kidney

C. renal medulla

39 Movement of substances from the peritubular capillary to the glomerular filtrate

H. tubular secretion

40 Outer region of the kidney — **G.** renal cortex

41 Vessel that conveys blood to the glomerulus — **F.** afferent arteriole

42 Anterior pituitary hormone that enhances water conservation — **D.** antidiuretic hormone

43 Movement of substances from the glomerular filtrate into the peritubular capillary — **J.** tubular reabsorption

44 Mass of cells located in the walls of glomerular arterioles that help regulate renal blood flow — **I.** juxtaglomerular apparatus

11 The immune and lymphatic systems

INTRODUCTION

The immune and lymphatic systems consist of a complex network of specialized cells and organs designed to protect and defend the body against attacks by pathogens. The immune system has two branches – innate and acquired. The innate system is in-built; the acquired mechanism involves the body's ability to develop resistance to specific diseases after exposure.

As blood passes through the capillaries, some of its plasma diffuses into the surrounding tissues, it is then known as interstitial fluid. Some fluid returns to the blood but some does not, instead it enters the lymphatic system becoming lymph.

The lymphatic system has three interrelated functions: (1) removal of interstitial fluid from tissues; (2) absorption and transportation of fats to the circulatory system; and (3) transportation of immune cells to and from the lymph nodes. Lymph flows through lymph nodes located along the lymphatic system. Cells of the lymph nodes ingest impurities, toxins and cellular waste. Lymph returns to the venous blood through lymphatic collecting ducts, the thoracic duct and the right lymphatic duct, entering the bloodstream at the right and left subclavian veins.

The paramedic has a key role in preventing and controlling infection, so it is essential to understand how the immune and lymphatic systems protect the body from pathogens that can cause illness or disease.

Useful resources

Paramedics! Test Yourself in Pathophysiology

Chapter 2

Anatomy and Physiology (8th edition)

Chapter 19

Cells of the immune system:

http://www.cliffsnotes.com/WileyCDA/CliffsReviewTopic/Cells-of-the-Immune-System.topicArticleId-8524,articleId-8475.html

For more information on the immune and lymphatic systems:

http://www.emc.maricopa.edu/faculty/farabee/BIOBK/BioBookIMMUN.html

---

LABELLING EXERCISE

1–12   Identify the organs and tissues of the lymphatic system in Figure
        11.1, using the options provided in the box on p. 173.

Figure 11.1 The organs and tissues of the lymphatic system

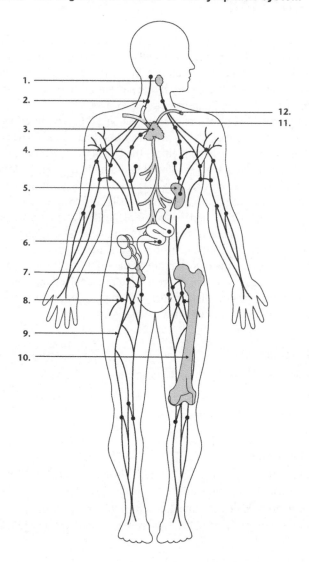

bone marrow   lymphatic vessels

tonsils   appendix

spleen   Peyer's patches

thymus gland   inguinal lymph nodes

sub-mandibular lymph nodes   axillary lymph nodes

thoracic duct   right lymphatic duct

TRUE OR FALSE?

Are the following statements true or false?

13 All cells are bathed in blood.

14 The heart pumps lymph around the body.

15 Lymph vessels contain one-way valves.

16 The innate immune response is non-specific.

17 In the body, white blood cells are more numerous than red blood cells.

18 Lymphocytes are formed in the bone marrow.

19 Antigens are found on the surface of cells.

20 Blood group O is called the universal recipient.

MULTIPLE CHOICE

Identify one correct answer for each of the following.

21 The immune system can be classified into two main categories:
a) humoral and cell-mediated
b) specific and non-specific
c) specific and humoral
d) cell-mediated and specific

22 The main function of the lymphatic system is:
a) to distribute hormones
b) to protect the body against external and internal threats
c) to remove damaged plasma proteins
d) to help maintain blood pressure

23 What is the first line of defence in any immune response?
a) anatomical, mechanical and chemical barriers
b) B- and T-lymphocytes
c) antibodies
d) macrophages

24 What is the large vessel running parallel to the spinal column transporting lymph called?
a) the aorta
b) the lymphatic duct
c) the thoracic duct
d) the right subclavian vein

25 Lymphadenopathy refers to:

a) an absence of lymph
b) lymph being found outside the vessels and ducts
c) chronic inflammation of the lymph nodes
d) surgical removal of certain lymph nodes

26 Which blood group can be transfused into an O Rh− patient?

a) O Rh−
b) O Rh+
c) AB Rh+
d) AB Rh−

27 Antibodies are formed from which white blood cell?

a) neutrophil
b) basophil
c) B-lymphocyte
d) monocyte

28 Identify the second of the four stages in an immune response:

a) plasma cells make antibodies
b) phagocytes engulf pathogens
c) T-cells activate B-cells to produce plasma cells
d) phagocytes activate T-cells

29 Compared to a primary immune response, a secondary response is described as being:

a) slower and less intense
b) faster but less intense
c) slower but more intense
d) faster and more intense

FILL IN THE BLANKS

Fill in the blanks in each statement using the options in the box below.
Not all of them are required, so choose carefully!

immunization (or vaccination)	agglutinate (coagulate, cross-react)
innate	Rhesus (Rh)
complement system	transfuse
thicken	inflammation
passive	humoral
antigen	

30 _____ is one of the first responses to infection or irritation.

31 The production of antibodies from B-cells is the _____ branch of the acquired (specific) immune response.

32 Artificial acquired immunity is achieved by _____.

33 The _____ _____ is a chemical part of the innate, humoral immune response.

34 The alternative to active immunity is _____ immunity.

35 If an A Rh+ patient is transfused with B Rh+ blood, the patient's blood will _____.

36 _____ factor is an antigen that may be present on the surface of blood cells.

ANSWERS

LABELLING EXERCISE

Figure 11.2 The organs and tissues of the lymphatic system

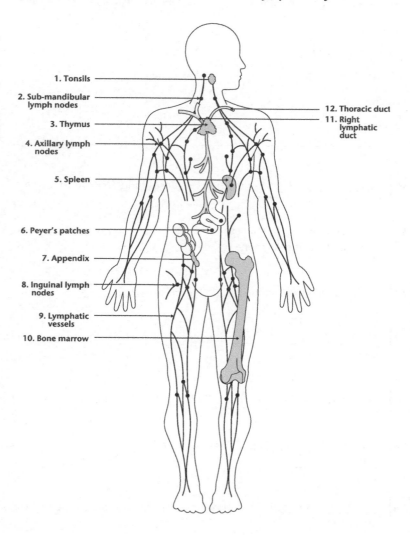

1. Tonsils
2. Sub-mandibular lymph nodes
3. Thymus
4. Axillary lymph nodes
5. Spleen
6. Peyer's patches
7. Appendix
8. Inguinal lymph nodes
9. Lymphatic vessels
10. Bone marrow
12. Thoracic duct
11. Right lymphatic duct

1 | **Tonsils:** lymphatic nodules. The adenoids (pharyngeal tonsils) and palatine tonsils are part of the immune system protecting the body against invasion by foreign substances and produce antibodies and lymphocytes. The body can function normally even when the adenoids and tonsils are removed.

2 | **Sub-mandibular lymph nodes:** located beneath the mandible. Lymph nodes are small bodies located in clusters throughout the body along lymphatic vessels. As lymph passes though them, it is filtered and any foreign substances are removed and destroyed by phagocytosis or by the T-cells of the immune system.

3 | **Thymus gland:** immediately after birth the thymus gland starts producing the T-lymphocytes (T-cells) responsible for cellular immunity. After a few months, this is complete and the gland has no further function in the immune system. It begins to degenerate (waste away) with only a non-functioning remnant remaining in adulthood.

4 | **Axillary lymph nodes:** located in the armpit, are responsible for draining lymph from the arms and breasts.

5 | **Spleen:** the largest mass of lymphatic tissue in the body. It is a highly vascular organ and acts as a reservoir and filter of the blood (but not lymph, like other lymph nodes), releasing additional blood into the circulatory system as needed. It is also involved with destruction of old cells and other substances by phagocytosis. It is located on the left side of the abdomen behind the stomach and above the kidneys. It produces specific antibodies in response to antigens, lymphocytes and plasma cells. Lymphocytes and free antibodies are added to the blood when it passes through the spleen. The spleen produces erythrocytes (red blood cells) during foetal development.

6 | **Peyer's patches:** a cluster of lymph nodes in the wall of the ileum where they provide protection against pathogens in the GI tract and protect against absorption of toxins from the contents of the GI tract.

7 | **Appendix:** short, thin tube, 7–10 cm long and attached to the caecum. As with the tonsils and adenoids, the body can survive and function as normal without the appendix so it is classified as an accessory organ of the immune system.

8 | **Inguinal (femoral) lymph nodes:** drain lymph from the lower limbs.

9 | **Lymphatic vessels:** form a network of vessels to return lymph back to the blood. They are structurally similar to veins and also possess valves to prevent back-flow because, as with veins, the lymph travelling in these vessels is under low pressure.

10 **Bone marrow:** blood-producing tissue located in the medullary cavity of certain bones. Stem cells give rise to all of the different types of blood cells (red and white) during haematopoiesis. There are two types of bone marrow – red and yellow. Red marrow consists of myeloid tissue for haematopoiesis while yellow marrow consists mainly of fat cells. In children, all bone marrow is red and is involved in forming blood cells. In adults, blood cells are only manufactured in red bone marrow which is found exclusively in the bones of the trunk and skull. The bones of the limbs contain yellow marrow which is not involved in making blood cells but can be converted to red marrow when more blood is needed.

11 **Right lymphatic duct:** receives lymph from the right side of the head, upper right trunk and right arm. It returns the lymph back into the systemic blood circulation at the junction of the right subclavian and right jugular veins.

12 **Thoracic duct:** (sometimes called left lymphatic duct) the most important of the lymphatic vessels. It receives lymph drained from all areas except the regions that drain to the right lymphatic duct and returns it to the blood circulation at the left subclavian and left jugular veins.

TRUE OR FALSE?

13 **All cells are bathed in blood.**

All cells are bathed in interstitial (tissue/extracellular) fluid. The circulating blood supplies nutrients and oxygen to the cells and tissues, and receives the products that cells produce. This requires exchange of substances between the blood and tissues. However, this exchange is not direct; it occurs through an intermediary called interstitial fluid. Interstitial fluid occupies the spaces between the cells and is their immediate environment. Interstitial fluid is formed from blood plasma that diffuses from the arteriole-end of capillaries. It has a milky/clear colour but unlike blood, interstitial fluid does not contain red blood cells or big proteins because both are too large to cross the capillary walls (unless the capillaries are damaged). Some interstitial fluid returns to the capillaries at the venule end; any fluid that remains is drained into blind-ending, thin-walled vessels called lymphatic vessels. The fluid is then called lymph.

14 **The heart pumps lymph around the body.**

The main difference between the blood flowing in the circulatory system and lymph flowing in the lymphatic system is that blood is pressurized by the heart, while the lymphatic system is not pressurized. While the heart acts as a central pump for the blood, there is no pump for lymph in the lymphatic system. Instead, fluid oozes into the lymphatic system and moves into the lymph nodes by normal skeletal and muscle motion.

15 **Lymph vessels contain one-way valves.**

Like the blood in the veins, lymph flows through the lymphatic system under very little pressure. Therefore to prevent back-flow, the vessels of the lymphatic system possess valves.

16 **The innate immune response is non-specific.**

This means that the innate immune system recognizes and responds to pathogens in a generic way, but unlike the acquired immune system, the innate system does not confer long-lasting or protective immunity to the host. Innate immunity has a number of characteristics: (1) responses are broad-spectrum (that is, non-specific); (2) there is no memory or lasting protective immunity; and (3) the number of recognition molecules is limited so it is not always sufficient as a defence mechanism. Potential pathogens are routinely encountered, but only rarely cause disease. The vast majority of microorganisms are destroyed within minutes or hours by the innate immune system. The acquired immune response becomes involved only when innate defences are breached and not sufficient to overcome the invading pathogen. Innate immunity is found in nearly all forms of life.

17 **In the body, white blood cells are more numerous than red blood cells.**

White blood cells (WBCs, leucocytes) are larger than red blood cells but they are significantly fewer. Typically, they are rounded cells that possess a nucleus and cytoplasm. WBCs are classified as granular or agranular depending on their appearance when viewed through a microscope. When the cytoplasm contains many granules and their nuclei appear to have many lobes, the cells are classified as granular (granulocytes). Within the granulocyte family there are a number of different WBCs: neutrophils, eosinophils and basophils. Some WBCs do not have granular cytoplasm and therefore are described as agranular (non-granular). The agranular WBCs are: monocytes and the B- and T-lymphocytes which are involved in the formation of antibodies for the immune response against foreign pathogens.

18 **Lymphocytes are formed in the bone marrow.**

All granular WBCs, along with the agranular monocytes (but not B- and T-lymphocytes), are manufactured in the red bone marrow while the B- and T-lymphocytes are formed in the lymphatic tissue from stem cells. When activated, the lymphocytes produce the immune cells that provide humoral immunity – these are the antibodies. Lymphocytes are involved in cell-mediated and humoral immune responses of the acquired immune system.

19 **Antigens are found on the surface of cells.**

Antigens are protein or glycoprotein molecules found on the surface membrane of cells. When a pathogen invades the body, it is the antigens on its cell surface that the body identifies as being foreign. This recognition then stimulates an immune response to eliminate the foreign invader.

20 **Blood group O is called the universal recipient.**

Blood group O is the universal donor. However, to be more specific, people who are blood group O Rh– are the true universal donors because they can donate blood to any blood group as they have no cell surface antigens *and* no Rhesus factor on the surface of their red blood cells. An O Rh+ donor could still trigger agglutination if transfused into a Rhesus negative patient. Conversely, blood group AB Rh+ is considered the universal recipient because these people have no antibodies and possess both A and B cell surface antigens and the Rhesus factor, therefore no blood type will trigger agglutination.

MULTIPLE CHOICE

Correct answers identified in bold italics.

21 The immune system can be classified into two main categories:

a) humoral and cell-mediated   ***b) specific and non-specific***
c) specific and humoral   d) cell-mediated and specific

The specific immune system is also known as the acquired (or adaptive) immune system, because acquired immunity is only obtained when an individual comes in contact with a certain pathogen and develops *specific* immunity to the pathogen. The non-specific immune system is also known as the innate immune system. Innate immunity is the unspecific responses that the body possesses to defend itself against pathogens. Innate immunity refers to a basic resistance against disease – the first line of defence against infection. The specific and non-specific defence systems can be divided into humoral and cellular (see Figure 11.3).

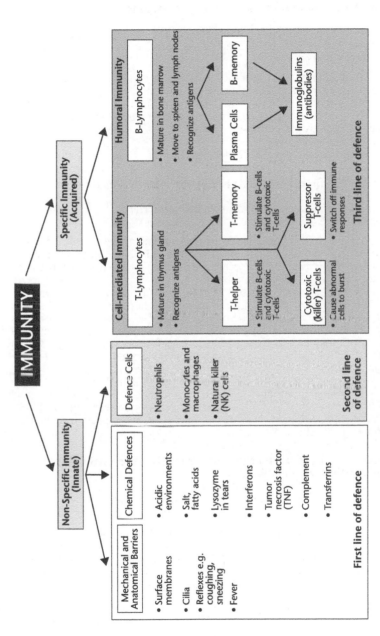

Figure 11.3 Summary of innate and acquired immune responses

22 | **The main function of the lymphatic system is:**

a) to distribute hormones   *b) to protect the body against external and internal threats*   c) to remove damaged plasma proteins   d) to help maintain blood pressure

The lymphatic system's primary function is to protect the body from infection and disease. These threats may come in the form of viruses which live within cells as they have no cellular structure themselves consisting only of proteins and nucleic acid. Bacteria and parasites also pose a threat to the human body which relies on the lymphatic system in part to assist in the defence and protection from the consequences of these internal and external pathogens.

23 | **What is the first line of defence in any immune response?**

a) anatomical, mechanical and chemical barriers   b) B- and T-lymphocytes   c) antibodies   d) macrophages

The first lines of immune defence are the anatomical and chemical barriers that external invaders will encounter. The skin is a major anatomical barrier and the first line of defence against infection. However, the body is not completely sealed against its environment; other systems act to protect body openings such as the lungs, intestines and the genitourinary tract. In the lungs, coughing and sneezing mechanically expel pathogens and other irritants from the respiratory tract. The flushing action of tears and urine also mechanically expels pathogens, while mucus secreted by the respiratory and GI tract serves to trap and prevent pathogens from further entering the body. Inflammation is another first line defence mechanism.

The body also has chemical barriers that act as a first line of defence against invading pathogens. These include the salt, fatty acids and acidic environments within the GI tract, lysozyme in tears, the anti-viral interferons, sweat and skin secretions and a system of proteins called the complement system.

24 | **What is the large vessel running parallel to the spinal column transporting lymph called?**

a) the aorta   b) the lymphatic duct   *c) the thoracic duct*   d) the right subclavian vein

The thoracic duct receives lymph drained from all parts of the body except for those serviced by the right lymphatic duct (the right side of the head, the right upper trunk and the right arm). Lymph travelling in the thoracic duct is returned to the blood circulation at the junction of the left subclavian vein and left jugular vein.

25 **Lymphadenopathy refers to:**

a) an absence of lymph   b) lymph being found outside the vessels and ducts   *c) chronic inflammation of the lymph nodes*   d) surgical removal of certain lymph nodes

Lymphadenopathy can arise from either local infections or systemic illness. In conditions that cause generalized lymphadenopathy the liver and spleen can also be enlarged. It is worthy of note that the presence of enlarged lymph nodes spans a range of illnesses from relatively benign streptococcal throat infections to more sinister malignant conditions. The age of the individual is also an important consideration when faced with lymphadenopathy. In childhood weak antigenic stimuli can cause significant enlargement to the nodes and lymphoid tissue, whereas adults will rarely show such a generalized response.

Information from the clinical history is invaluable in the diagnostic management of the patient with lymphadenopathy, and frequently leads to an accurate diagnosis without the need for extensive diagnostic testing. The age of the patient is quite important. Dramatic enlargement of lymph nodes and other lymphoid tissue such as the adenoids and tonsils is often a normal response to a variety of relatively weak antigenic stimuli such as mild viral and bacterial infections or vaccinations in infants and children, whereas in adults these antigens will not elicit a generalized response. This age difference in the expression of lymphadenopathy is of such importance as to warrant an almost totally different diagnostic approach to patients before and after puberty.

26 **Which blood group can be transfused into an O Rh− patient?**

a) O Rh−   b) O Rh+   c) AB Rh+   d) AB Rh−

This patient can only receive precisely matched blood from an O Rh− donor because their O Rh− blood will possess both antibody 'a' and antibody 'b'. The presence of these antibodies means blood will react against both A and B antigens that would be present in A, B or AB blood groups. The O Rh− patient will also have no Rhesus antibodies and therefore will react to any blood that is Rh+ (even O Rh+). It is therefore essential that the patient receives precisely matched O Rh− blood in a transfusion otherwise agglutination (coagulation or cross-reaction) will occur which can have fatal consequences (see Figure 11.4).

Figure 11.4 The ABO blood groups

	Group A	Group B	Group AB	Group O
Cell-surface antigens present	A	B	A and B	NO cell surface antigens
Antibodies in plasma	Anti-b	Anti-a	NO ANTIBODIES	Anti-a and Anti-b
Safe donors *(incompatible donors cause agglutination)*	A O	B O	AB A B O	O (only)

---

27 **Antibodies are formed from which white blood cell?**

a) neutrophil   b) basophil   *c) B-lymphocyte*   d) monocyte

There are two types of lymphocyte – T-lymphocyte (T-cell) and B-lymphocyte (B-cell). The type of lymphocyte is determined by the location of cell maturation. All lymphocytes are formed from stem cells in either the thymus gland or lymphoid tissue. B-lymphocytes complete their development in the spleen or lymph nodes. Lymphocytes that mature in the thymus are referred to as T-lymphocytes (or T-cells). Within a few months of birth, the T-cells are fully mature and leave the thymus to circulate in the blood and body fluids.

28 **Identify the second of the four stages in an immune response:**

a) plasma cells make antibodies   b) phagocytes engulf pathogens
c) T-cells activate B-cells to produce plasma cells   *d) phagocytes activate T-cells*

The immune response can be divided into four stages:

I.   The first involves the engulfing of pathogens by phagocytes – making the pathogen visible for targeting by the immune response.

II. This visibility allows the T-cell receptors to bind to the foreign antigen.

III. Activated T-cells release substances to activate B-cells and stimulate the production of plasma cells and memory cells.

IV. Plasma cells produce many clones of the B-cell which secrete antibodies specific to the foreign antigen. These antibodies help to clear the infection.

29 **Compared to a primary immune response, a secondary response is described as being:**

a) slower and less intense   b) faster but less intense   c) slower but more intense   *d) faster and more intense*

When a foreign (non-self) antigen enters the body for the first time, it stimulates an immune response – this is the primary response. Since the body has never encountered the antigen before, it is slow to respond. However, during the primary response, the body also makes memory cells that it can use if it encounters the foreign pathogen in the future. When the body later encounters the same pathogen, the secondary immune response is much faster and more intense than the primary response because the memory cells exist and the body can react to the foreign antigen much quicker (see Figure 11.5). Often, in a secondary immune response, the body will have eradicated the pathogen before any symptoms develop.

Figure 11.5 Primary and secondary immune responses

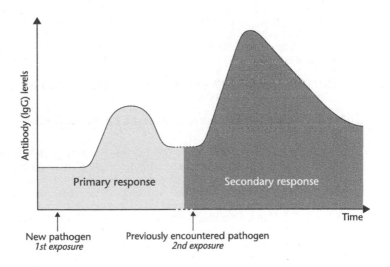

FILL IN THE BLANKS

30 *Inflammation* **is one of the first responses to infection or irritation.**

Inflammation describes a generic response that is not specific to any invading pathogen. It is characterized by the local accumulation of fluid, plasma proteins and white blood cells. Inflammation is initiated by physical injury, infection or a local immune response and is stimulated by chemical factors released by injured cells. It establishes a physical barrier preventing the spread of infection and promoting healing of damaged tissue following removal of pathogens. Chemical factors produced during inflammation sensitize pain receptors and cause vasodilation of the blood vessels at the scene. Mast cells release histamine which causes itching and swelling. The inflammatory response is characterized by the following symptoms: redness, heat, swelling, pain and possible loss of function of the organs or tissues involved.

31 **The production of antibodies from B-cells is the *humoral* branch of the acquired (specific) immune response.**

In the humoral (antibody-mediated) immune response, an invading antigen causes the B-cells to divide and differentiate into either plasma or memory cells. Each plasma cell secretes antibodies (immunoglobulins) into the blood which can detect the antigen on the surface of a foreign invader. An antibody is a specialized protein substance produced by the body in response to a foreign antigen. Antibodies are capable of reacting specifically with the antigen that provoked their production. Other activated B-cells form memory cells. These can be activated later for rapid antibody production when the foreign antigen is detected again and this is the basis of long-term immunity.

32 **Artificial acquired immunity is achieved by *immunization* (or *vaccination*).**

When a vaccine is injected into a healthy body, it stimulates the immune system to produce antibodies and memory cells to the antigens in the vaccine. Hence the body has acquired immunity to future infections with the same antigens – in effect, the vaccination stimulated a primary immune response so that the body now has the memory cells required to initiate a future secondary immune response. The body can initiate its own acquired immunity if it is exposed to an infecting pathogen because it produces its own antibodies to counteract the infection. In response to a subsequent infection, the body will launch a rapid, intense secondary response to the invading pathogen, hence the disease is not given the chance to fully develop. This is why people usually only contract chickenpox once in a lifetime.

33 | **The *complement system* is a chemical part of the innate, humoral immune response.**

The complement system is a biochemical catalytic cascade that attacks the surfaces of foreign invading cells. It links the innate and acquired immune systems by (1) enhancing antibody responses and immune 'memory'; (2) lysing foreign cells; and (3) removing antibody-antigen (immune) complexes and apoptotic cells. It contains 20–25 different enzymes and is named because of its ability to 'complement' (enhance) the antibodies' destruction of pathogens.

34 | **The alternative to active immunity is *passive* immunity.**

Passive immunity is provided when the body is given antibodies rather than producing them itself. A newborn baby has passive immunity to several diseases from antibodies passed from its mother via the placenta and breast milk. Passive immunity only lasts for a few weeks or months; usually this is long enough for the baby's immune system to develop. The short-term nature of passive immunity is also why babies are immunized against a number of diseases after the first few months of life – they need to develop their own antibodies. Another example of passive immunity is the anti-venom injection administered after certain snake bites.

35 | **If an A Rh+ patient is transfused with B Rh+ blood, the patient's blood will *agglutinate (coagulate, cross-react)*.**

If a person is transfused with non-compatible blood, agglutination will occur. This is triggered when an antibody and its specific antigen come together (see Figure 11.4 for antigens and antibodies). For example, when a patient who is blood group A is transfused with type B blood, the A antigens on the surface of cells of the type A blood react with 'a' antibodies produced by type B blood. This agglutination reaction causes the cells to clot, forming clumps of blood that can block small blood vessels and damage vital organs, such as the heart or brain, which can be fatal.

36 | ***Rhesus (Rh)* factor is an antigen that may be present on the surface of blood cells.**

Sometimes called Rhesus factor D, it is an antigen that is present or can be synthesized by the individual. About 85 per cent of the world's population are Rhesus positive (Rh+) meaning they possess the ability to manufacture the Rh antigen. In Rh negative (Rh–) individuals, their immune system will manufacture antibodies against the Rh antigen when they encounter it. Along with the ABO blood grouping, the Rhesus factor should always be considered when blood typing for transfusion. The Rhesus factor should be checked during pregnancy. A mother who is Rh– but carrying a Rh+ foetus may become sensitized to the Rhesus antigen and generate anti-Rhesus antibodies. This can threaten a subsequent pregnancy with Rh+ foetus. Nowadays this condition is well understood and is usually easily treated.

12 The male and female reproductive systems

INTRODUCTION

The organs of the male and female reproductive systems are considered unique to each sex. Nevertheless, a common general structure and function can be identified, which contribute to successful reproduction.

The male and female reproductive organs are adapted for the development of sperm and ova, followed by successful fertilization, embryogenesis and foetal development. Furthermore, the hormones that promote development of secondary sex characteristics are stimulated by the reproductive systems.

Sperm are the male gametes while female gametes are called oocytes. Each have one set of genetic information carried on 23 chromosomes (compared to two sets on 46 chromosomes in other body cells). When sperm and ovum combine at fertilization, the full amount of genetic material is usually available on the 46 chromosomes in the developing zygote.

For paramedics, understanding the reproductive systems is important when dealing not only with prepubescent patients (and the influence of hormones), but also when considering the functioning of the reproductive systems. One of the main problems associated with both the male and female reproductive systems is the growing rise in sexually transmitted infections, which may lead to infertility and other potentially serious illnesses.

Useful resources

Paramedics! Test Yourself in Pathophysiology

Chapter 11

Anatomy and Physiology (8th edition)

Chapter 24

A series of short videos on cell division and reproduction:

http://www.pbs.org/wgbh/nova/miracle/program.html

LABELLING EXERCISE

1–15   Identify the regions and structures of the male and female reproductive systems in Figure 12.1, using the options provided in the box on p. 192.

Figure 12.1 The female and male reproductive systems

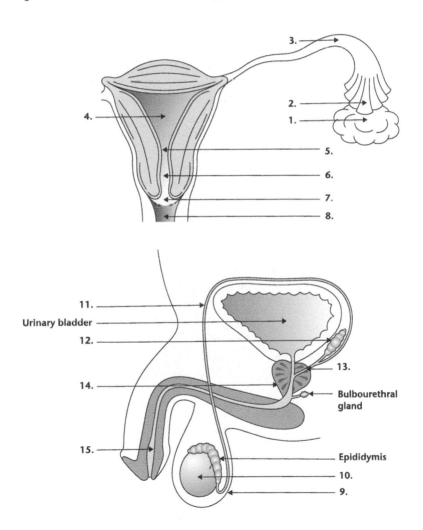

urethra   scrotum

cervical orifice   prostate gland

seminal vesicle   testis

Fallopian tube   ovary

vagina   ejaculatory duct

fimbriae   vas deferens

internal os   cervical canal

uterine cavity

TRUE OR FALSE?

Are the following statements true or false?

16 The sole function of the ovary is oogenesis.

17 Oestrogen stimulates development and maintenance of the female reproductive structures.

18 The uterine and ovarian cycles are controlled by the hormone oestrogen.

19 Ovulation occurs on day 0 of a 28-day female reproductive cycle.

20 Cervical mucus protects sperm from phagocytes.

21 In males, inhibin is produced by the Sertoli cells.

22 Mature sperm are stored in the bulbourethral glands.

23 The seminal vesicles secrete an acidic fluid that constitutes approximately 15 per cent of semen volume.

24 Oxytocin creates a positive feedback loop.

MULTIPLE CHOICE

Identify one correct answer for each of the following.

25 In males, testosterone is produced by:

a) the interstitial cells

b) the seminiferous tubules

c) the epididymis

d) the vas deferens

26 The male gonads are called:

a) the testes

b) the ovaries

c) the accessory sex glands

d) the sperm ducts

27 The accessory sex glands of the male include:

a) the prostate and paraurethral glands

b) the prostate and bulbourethral glands

c) the prostate and vestibular glands

d) the paraurethral and bulbourethral glands

28 The acrosome of sperm contains:

a) mitochondria

b) lysosomal enzymes

c) nuclear material

d) all of the above

29 A fertilized ovum is called:

a) a blastocyst

b) a secondary oocyte

c) a diploid cell

d) a zygote

30 The female external genitalia are called:

a) the pubic symphysis

b) the vagina

c) the vulva

d) the clitoris

31 Where does fertilization of an ovum by sperm usually take place?

a) in the Fallopian tube

b) in the vagina

c) in the uterus

d) in the ovary

32 Which hormone, produced by the ovaries, inhibits secretion of FSH and LH?

a) progesterone

b) oestrogen

c) inhibin

d) GnRH

33 The onset of reproductive age is termed:

a) menstruation

b) menarche

c) menopause

d) puberty

34 Progesterone is secreted by:

a) the ovarian follicle

b) the mature follicle

c) the corpus luteum

d) the corpus albicans

MATCH THE TERMS

Match each statement with the correct description.

A. ovulation   **D.** spermatogenesis

B. menopause   **E.** meiosis

C. contraception   **F.** menstruation

35 Development of immature sperm cells to mature sperm cells _____

36 Release of an ovum into the Fallopian tube _____

37 A form of cell division in which the number of chromosomes is halved in the daughter cells _____

38 Monthly discharge of blood, mucus, epithelial cells and tissue fluid _____

39 The permanent cessation of reproductive fertility _____

40 Birth control _____

ANSWERS

 LABELLING EXERCISE

Figure 12.2 The female and male reproductive systems

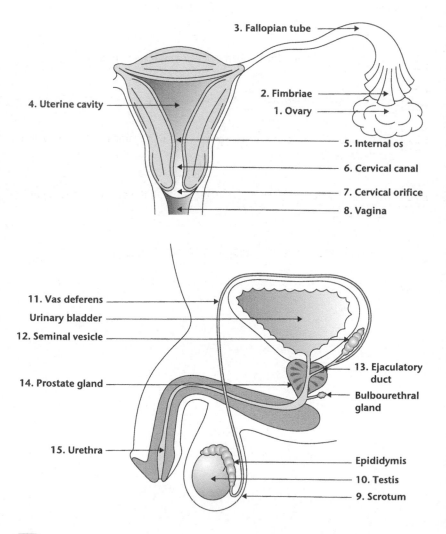

1 | **Ovary:** one of a pair of glands (pl. ovaries) in the female homologous to the testes in the male. The ovaries lie in the upper part of the pelvic cavity, one on each side of the uterus and held in place by ligaments.

2 | **Fimbriae:** a fringe of finger-like projections around the end of the Fallopian tube at the ovary. An ovary is not directly connected to its Fallopian tube. When ovulation is about to occur, the fimbriae brush the ovary in a gentle, sweeping motion. An oocyte is released from the ovary and the cilia of the fimbriae sweep it into the Fallopian tube.

3 | **Fallopian tube (oviduct, uterine tube):** one of a pair of tubes, about 10 cm long, which extend away from the uterus and transport the ova from the ovaries to the uterus. The ova are conveyed along the tube by peristalsis of the muscle layer and by the sweeping action of cilia lining the tube.

4 | **Uterine cavity (uterus):** small, pear-shaped muscular organ located between the bladder and rectum that receives an embryo and sustains its development. The mucous membrane lining the uterus is called the endometrium and has a rich blood supply.

5 | **Internal os:** the narrow upper opening between the cervix and the uterine cavity.

6 | **Cervical canal:** a spindle-shaped, flattened canal of the cervix that connects the uterine cavity with the vagina.

7 | **Cervical orifice (external os):** small, circular opening between the cervical cavity and the vagina.

8 | **Vagina:** elastic muscular canal approximately 10 cm long, extending from the cervix to the vulva. It provides a path for menstrual blood and tissue and also as a delivery channel during birth. During sexual arousal the vagina expands in both length and width. Its elasticity allows it to stretch during sexual intercourse and during the birth process. The vagina connects the superficial vulva to the cervix of the deep uterus. Before and during ovulation, cervical mucus glands secrete different variations and consistencies of mucus, which provide a favourable alkaline environment within the vaginal canal for sperm survival.

9 | **Scrotum:** a sac that hangs from the base of the penis. The scrotum is the supporting structure for the testes. Internally, it divides into two sacs, each containing one testis. The location of the scrotum and the contraction of its muscle fibres regulate the temperature of the testes. Both production and survival of sperm require a temperature that is slightly lower than core body temperature.

10 **Testis:** one of a pair of oval glands (pl. testes) in the scrotum cavity suspended by the spermatic cords. The testes are the male organs of reproduction. They contain the seminiferous tubules that produce sperm cells and the male sex hormone, testosterone (produced by interstitial (Leydig's) cells within the seminiferous tubules).

11 **Vas deferens (sperm ducts):** a pair of ducts connecting the left and right epididymis to the ejaculatory ducts, facilitating movement of sperm. Each duct is about 30 cm long and surrounded by a layer of smooth muscle. The procedure of vasectomy is a method of male sterilization in which these ducts are permanently cut, although in some cases it can be reversed.

12 **Seminal vesicle:** a pair of accessory glands which produce a viscous alkaline fluid, seminal fluid, that neutralizes acid in the female reproductive tract, which would otherwise inactivate and decrease sperm viability. It contains a number of components such as fructose, prostaglandins and proteins that coagulate sperm, contributing to its motility and viability.

13 **Ejaculatory duct:** a duct formed by the joining of the vas deferens and the duct from the seminal vesicle that allows sperm to enter the prostatic urethra prior to ejaculation. It also transports and ejects the secretions of the seminal vesicles.

14 **Prostate gland:** lying below the bladder and surrounding part of the urethra, this gland secretes a slightly alkaline fluid which contains a number of enzymes that dissolves coagulated semen. Secretions of the prostate gland enter the prostatic urethra through a number of prostatic ducts. These secretions contribute approximately 25 per cent of the total semen volume and promote sperm motility and viability.

15 **Urethra:** a duct leading from the urinary bladder to the exterior of the body that conveys urine in females and both urine and semen in males.

TRUE OR FALSE?

16 **The sole function of the ovary is oogenesis.**

Oogenesis is the production of the female gametes, the ova (eggs). Additional functions of the ovaries include the production of the hormones oestrogen, progesterone, relaxin and inhibin.

17 **Oestrogen stimulates development and maintenance of the female reproductive structures.**

Oestrogen also stimulates development of secondary sex characteristics and breast tissue and helps regulate fluid and electrolyte balance and increase protein metabolism. It also assists in reducing blood cholesterol levels. Moderate levels of oestrogen in the blood inhibit the release of gonadotropin-releasing hormone (GnRH) by the hypothalamus and the secretion of luteinizing hormone (LH) and follicle-stimulating hormone (FSH) by the anterior pituitary. This inhibition forms the basis for the pharmacological action of oral contraception.

18 **The uterine and ovarian cycles are controlled by the hormone oestrogen.**

The uterine and ovarian cycles are controlled by GnRH from the hypothalamus. GnRH stimulates the release of FSH and LH from the anterior pituitary gland. FSH stimulates the initial secretion of oestrogen by growing follicles in the ovary. LH stimulates the further development of ovarian follicles and their full secretion of oestrogen. This initiates ovulation, promotes corpus luteum formation from the ruptured follicle after ovulation and induces the production of oestrogen, progesterone, relaxin and inhibin.

19 **Ovulation occurs on day 0 of a 28-day female reproductive cycle.**

The average length of the female reproductive cycle is 28 days. It can be divided into three phases: (1) the menstrual phase; (2) the proliferative (pre-ovulatory) phase; and (3) the secretory (post-ovulatory) phase. The menstrual phase lasts for approximately the first five days of the cycle. By convention the first day of menstruation marks the beginning of a new cycle. The second phase is the time between menstruation and ovulation. This proliferative phase of the cycle is more variable than the other phases, it typically lasts from days 6 to 13 in a 28-day cycle. Ovulation occurs on day 14. The secretory phase is the most constant in terms of duration and lasts for 14 days. It represents the time between ovulation and the onset of the next menses.

20 **Cervical mucus protects sperm from phagocytes.**

The secretory cells in the mucosa of the cervix produce a secretion called cervical mucus. It is a mixture of water, glycoprotein, serum-type proteins, lipids, enzymes and inorganic salts. Females of reproductive age secrete 20–60 mL of cervical mucus each day. The mucus is more receptive to sperm around the time of ovulation since it is then less viscous and more alkaline (pH 8.5). At other times the mucus forms a plug that physically impedes sperm penetration. The mucus also provides nutrients for sperm metabolism.

21   **In the male, inhibin is produced by the Sertoli cells.**

Sertoli (or nurse) cells secrete a range of substances related to reproductive function. The main function of Sertoli cells is to nurture developing sperm cells through the stages of spermatogenesis. The aromatase enzyme from Sertoli cells is involved in the conversion of androgens to oestrogens which help regulate spermatogenesis.

22   **Mature sperm are stored in the bulbourethral glands.**

Mature sperm are stored in the vas deferens where they can remain viable for many months. From here, sperm is conveyed from the epididymis toward the urethra by muscular peristalsis. Sperm cells that do not ejaculate are ultimately reabsorbed. The bulbourethral glands are a pair of glands lying below the prostate gland on either side of the urethra. During sexual arousal the bulbourethral glands secrete an alkaline substance that protects sperm.

23   **The seminal vesicles secrete an acidic fluid that constitutes approximately 15 per cent of semen volume.**

The seminal fluid (semen) secreted by the seminal vesicles is alkaline and contributes around 60 per cent of semen volume. It is rich in fructose for energy production and prostaglandins to assist in sperm motility and viability. Prostaglandins may also stimulate muscular contraction within the female reproductive tract which also promotes sperm motility.

24   **Oxytocin creates a positive feedback loop.**

Oxytocin is one of the few hormones to create a positive feedback loop. During labour, uterine contractions stimulate the release of oxytocin from the posterior pituitary which, in turn, increases uterine contractions. This positive feedback loop continues throughout the labour process and ceases with birth.

a b c d   MULTIPLE CHOICE

Correct answers identified in bold italics.

25   **In males, testosterone is produced by:**

a) the interstitial cells   b) the seminiferous tubules
c) the epididymis   d) the vas deferens

Interstitial (Leydig) cells are located between the seminiferous tubules in the testes. They secrete testosterone, the most important of the androgen hormones in males, when stimulated by the LH from the pituitary gland. The seminiferous tubules are tightly coiled ducts where sperm cells are produced. The epididymis is the site of sperm maturation and the vas deferens is the duct which carries sperm from the epididymis to the ejaculatory duct.

26 **The male gonads are called:**

a) the testes   b) the ovaries   c) the accessory sex glands   d) the sperm ducts

The gonads are glands that produce gametes and sex-related hormones. In males, these are the paired testes and in the female, the ovaries. In males, the accessory sex glands include the seminal vesicles, the prostate gland and the bulbourethral glands. All are involved in the production of the liquid portion of semen.

27 **The accessory sex glands of the male include:**

a) the prostate and paraurethral glands   *b) the prostate and bulbourethral glands*   c) the prostate and vestibular glands   d) the paraurethral and bulbourethral glands

The male accessory sex glands are the prostate gland, the bulbourethral glands and the seminal vesicles. They secrete the liquid portion of semen whereas the ducts of the male reproductive system store and transport sperm cells. The paraurethral gland (sometimes called the 'female prostate' because it is similar to the male prostate) produces and secretes mucus in females. The vestibular glands are found on either side of the vaginal opening (vaginal orifice) and produce mucus secretions during sexual arousal that, along with cervical mucus, provide lubrication.

28 **The acrosome of sperm contains:**

a) mitochondria   *b) lysosomal enzymes*   c) nuclear material   d) all of the above

The acrosome is a cap-like structure on the head of sperm cells. Acrosome formation is completed during testicular maturation of the sperm. The acrosome contains a number of digestive enzymes which break down the outer membrane of the ovum, allowing the 23 chromosomes in the sperm to join with the 23 chromosomes in the ovum forming a diploid zygote with all 46 chromosomes.

29 **A fertilized ovum is called:**

a) a blastocyst   b) a secondary oocyte   c) a diploid cell   *d) a zygote*

A zygote is the single diploid cell, or fertilized ovum, which results from the fusion of the male and female gametes. A blastocyst describes the early development of an embryo as a hollow ball of dividing cells. A secondary oocyte is formed during oogenesis and involves several phases including meiosis. A diploid cell contains two sets of chromosomes, one set from each parent. Human somatic cells have 23 pairs of chromosomes, therefore diploid cells have 46 chromosomes. Gametes are not diploid cells (they are haploid) because they have only one set of chromosomes.

30 **The female external genitalia are called:**

a) the pubic symphysis   b) the vagina   *c) the vulva*   d) the clitoris

The vulva is the collective name for the external female genitalia which consists of the mons pubis, labia majora, labia minora, clitoris and vestibule. The mons pubis is adipose tissue, covered by skin and pubic hair that cushions the pubic symphysis. The labia majora are two folds of skin, homologous to the scrotum. The labia minora are between the labia majora but have no adipose tissue or pubic hair, but do possess numerous sebaceous glands. The clitoris is a small, cylindrical mass of erectile tissue and nerves, it is homologous to the male penis because it contains erectile tissue and many nerve endings. The vestibule describes the space between the labia minora.

31 **Where does fertilization of an ovum by sperm usually take place?**

a) in the Fallopian tube   b) in the vagina   c) in the uterus
d) in the ovary

Fertilization usually occurs in the Fallopian tube, 12–24 hours after ovulation. Since ejaculated sperm remain viable for approximately 48 hours and an ovum is viable for about 24 hours after ovulation, there is a 72-hour window during which fertilization can occur. After several days of cell division a blastocyst develops which eventually attaches to the uterine wall. Sometimes an embryo may develop outside the uterus either in the pelvic cavity, or more commonly in the Fallopian tube, this is an ectopic pregnancy. Unless removed, the developing embryo will result in a medical emergency.

32 **Which hormone, produced by the ovaries, inhibits secretion of FSH and LH?**

a) progesterone   b) oestrogen   *c) inhibin*   d) GnRH

Ovaries produce inhibin which inhibits secretion of FSH and LH from the anterior pituitary. During pregnancy, the ovaries and the placenta produce the hormone relaxin, which increases the flexibility of the pubic symphysis during pregnancy and induces dilation of the uterine cervix during labour and delivery.

33 **The onset of reproductive age is termed:**

a) menstruation   b) menarche   c) menopause   *d) puberty*

During puberty the secondary sex characteristics begin to appear and sexual reproduction becomes possible. Menstruation describes the periodic discharge of blood, mucus and epithelial cells, and marks the beginning of a new menstrual cycle. Lasting approximately five days, it is triggered by a sudden reduction in oestrogen and progesterone levels. Menarche relates to the first menstrual flow and beginning of the reproductive cycles in the female. Menopause describes the cessation of these cycles and the end of the reproductive era.

34 **Progesterone is secreted by:**

a) the ovarian follicle   b) the mature follicle   *c) the corpus luteum*   d) the corpus albicans

Progesterone is secreted mainly by the cells of the corpus luteum which is the follicle tissue left behind in the ovary after the release of the ovum during ovulation. Progesterone acts synergistically with oestrogen to prepare the endometrium of the uterus for implantation of a blastocyst and the mammary glands for milk production. High levels of progesterone also inhibit the secretion of GnRH and luteinizing hormone (LH).

 MATCH THE TERMS

35 Development of immature sperm cells to mature sperm cells   **D.** spermatogenesis

Spermatogenesis describes the development of immature male sperm cells (spermatogonia) into mature sperm cells (spermatozoa). It is a stepwise process which takes place within several structures of the male reproductive system over approximately 65 days. The initial stages occur within seminiferous tubules of the testes and progress to the epididymis where the developing gametes mature and are stored until ejaculation. Spermatogenesis is highly dependent on optimal conditions for the process to occur correctly and is essential for sexual reproduction. It starts at puberty and usually continues uninterrupted throughout life although a slight decrease in the quantity of sperm produced is associated with increasing age.

36 Release of an ovum into the Fallopian tube   **A.** ovulation

Ovulation occurs when a mature follicle ruptures and releases an ovum into the Fallopian tube which is triggered by increasing LH levels. In a normal 28-day cycle this usually occurs around day 14. The remaining follicle develops into the corpus luteum which degenerates by day 28 if pregnancy does not occur.

37 A form of cell division in which the number of   **E. meiosis**
chromosomes is halved in the daughter cells

A process of cell division in which the number of chromosomes per cell is halved resulting in the formation of gametes (sex cells). As with mitosis, before meiosis begins, the DNA in the parent cell is replicated during the S-phase of the cell cycle. Two cell divisions separate the replicated chromosomes into four haploid gametes. Each of these gametes contains one complete set of chromosomes, that is, half the genetic content of the parent cell. These haploid gametes must fuse during fertilization to create a new diploid cell, or zygote.

38 Monthly discharge of blood, mucus, epithelial   **F. menstruation**
cells and tissue fluid

This periodic discharge of blood, tissue fluid, mucus and epithelial cells is caused by a reduction in oestrogens and progesterone. It occurs each month after puberty until the menopause with interruptions associated with pregnancy. Regular menstruation typically lasts 3–5 days with an average blood loss of 35 mL. Due to this blood loss, premenopausal women require higher amounts of iron in their diet to prevent deficiency. Many women experience uterine cramps (dysmenorrhoea) during this time, due to contractions of the uterine muscle as it expels the endometrial blood and tissue.

39 The permanent cessation of reproductive   **B. menopause**
fertility

Menopause is part of a biological process that most women first notice in their mid-forties. During this transition, the ovaries start producing lower levels of oestrogen and progesterone. Oestrogen promotes the normal development of breast and uterine tissue, controls the ovulation cycle and affects many aspects of physical and emotional health. Progesterone controls menstruation and prepares the lining of the uterus to receive the fertilized egg. With a marked reduction in the production of these hormones, reproduction becomes impossible.

40 Birth control   **C. contraception**

Contraception is the general term for the various types of birth control. Although there is no single, ideal method of birth control, several methods exist, each with advantages and disadvantages. The most popular methods in the UK include sterilization, hormonal methods (such as the contraceptive pill), intrauterine devices (IUD), barrier and chemical methods. Physiological methods based on knowledge of the body changes that occur during the menstrual cycle are popular in some societies although they have higher failure rates. Certain barrier contraceptives (male/female condom, cervical cap, diaphragm) have additional health benefits by reducing the transmission of sexually transmitted infections.

Glossary

Abdomen: area of the body between the diaphragm and pelvis.

Adenoids: pair of lymphoid structures located in the nasopharynx.

Adrenal glands: two secretory organs that sit on the kidneys.

Alveolus: (pl. alveoli) small grape-like structure located at the terminus of the bronchioles. The site of gas exchange in the lungs.

Antibody: immunoglobulin produced by the body in response to exposure to a specific antigen.

Antidiuretic hormone (ADH): or vasopressin; regulates the body's conservation of water. It is secreted from the posterior pituitary gland when the body is dehydrated causing the kidneys to reabsorb more water, producing a lower volume but more concentrated urine.

Antigen: structure detected on the surface of a foreign cell that has entered the body, it stimulates formation of antibodies by the body.

Anus: outlet of the rectum.

Anvil: or incus, anvil-shaped bone which is one of three bones in the middle ear.

Aorta: main blood vessel of the arterial circulation. It originates from the left ventricle of the heart and branches into a number of main arteries that serve the systemic circulation.

Apex: top, tip or pointed end of an organ.

Apoptosis: a normal part of cell development when the cell undergoes programmed cell death through a series of biochemical changes in the cell. The cell debris from apoptosis does not cause harm to adjacent cells, this is how it differs from necrosis.

Arachnoid: delicate middle membrane of the meninges in the brain and spinal cord.

Arteriole: smaller branch of an artery.

Atrium: (pl. atria) chamber of the heart; sits on top of the ventricles and is smaller in size.

Auricle: exterior portion of the ear that is attached to the head.

Auscultation: listening to internal sounds of body, usually with a stethoscope.

Axon: long extension of a neurone that conveys electrical nerve impulses away from the cell body towards the next neurone.

Baroreceptors: detect the blood pressure as blood passes through them, send messages to the CNS to increase or decrease cardiac output and total peripheral resistance.

Basal ganglia: area of brain that modifies and coordinates voluntary muscle movements.

Bile: green-yellow alkaline fluid secreted from the liver and stored in the gall bladder. It aids digestion by emulsifying fats in the duodenum.

Bladder: muscular sac that holds waste liquid before excretion from the body during urination.

Blood pressure: force exerted by the circulating blood on walls of blood vessels.

Bone: hard, dense connective tissue of the skeleton.

Bone marrow: soft tissue located in the spongy (or trabecular) bone of the epiphysis; blood cell formation and maturation occur here.

Brain stem: area of brain housing the pons, medulla oblongata and midbrain.

Broca's area: region of brain, within the cerebrum associated with speech motor function.

Bronchiole: small branch of the bronchus.

Bronchus: larger air passage of the lungs.

Buccal: referring to the cheek.

Caecum: pouch located at the proximal end of the large intestine, next to the appendix.

Capillary: very small blood vessel branching from the arterioles and linking with venules of the venous system. Exchange of cellular materials can only occur in the capillaries.

Carpal: refers to the wrist.

Cartilage: connective supporting tissue usually located in joints, the thorax, larynx, trachea, nose and ear.

Central nervous system (CNS): one of the two main divisions of the nervous system. Refers to the brain and spinal cord.

Cerebellum: small section of the brain located in the posterior region, behind the brain stem; it is responsible for coordinating voluntary muscular activity.

Cerebral cortex: outermost layer of grey matter covering the cerebrum of the brain, controls higher mental activities.

Cerebrum: largest and uppermost section of the brain, divided into two hemispheres, left and right (each is sub-divided into four lobes).

Cilia: small, hair-like projections on the surface of some cells, especially in respiratory tract.

Coagulation: clotting (especially of blood).

Cochlea: spiral tube located in the inner ear.

Coeliac: relating to the abdomen.

Colon: region of the large intestine extending from the caecum to the rectum.

Cornea: transparent portion of the eye, convex shape.

Coronary: referring to the heart or its vessels.

Cortex: outer part of an internal organ, opposite of medulla.

Costal: refers to the ribs.

Cranium: (skull) bone that encases and protects the brain.

Cutaneous: relating to the skin.

Deltoid: triangular in shape.

Dendrite: branch projection that extends from the cell body of a neurone, directs nerve impulse towards the cell body.

Dermis: skin layer located below the epidermis.

Diaphragm: muscular partition that separates the thorax from the abdomen.

Diaphysis: shaft portion of a long bone.

Diarthrosis: freely movable joint such as hip or shoulder.

Diastole: relaxation of heart muscle when chambers fill with blood; opposite of systole.

Diencephalon: part of the brain located between cerebral hemisphere and midbrain.

Duct: a canal or passage.

Duodenum: shortest and widest portion of the small intestine; extends from the pylorus of stomach to the jejunum.

Dura mater: outermost layer of the meninges surrounding the brain or spinal cord.

Ear: the organ for the sense of hearing.

Endocardium: interior lining of the heart.

Endocrine: refers to a secretion into the blood or lymph (rather than into duct); the opposite of exocrine. Referring to hormonal system.

Epidermis: outmost layer of skin, it lacks vessels.

Epiglottis: cartilaginous structure that overhangs the larynx preventing food from entering the respiratory tract.

Epiphyses: (sing. epiphysis) end portions of long bone.

Epithelium: tissue layer lining the exterior and interior of organs and cavities in the body.

Erythrocyte: red blood cell.

Exocrine: refers to a secretion into a duct (rather than into blood or lymph); the opposite of endocrine.

Extension: straightening limbs at a joint.

Eyes: the organs of vision.

Fallopian tubes: pair of ducts extending from each ovary to the uterus.

Femur: thigh bone. The longest and largest bone of the human anatomy.

Fibula: calf bone. Located on the lateral side of the tibia, with which it is connected although it is smaller than the tibia.

Flexion: bending the limbs at a joint.

Fossa: hollow region or cavity.

Fundus: base of a hollow organ, region furthest away from the organ's outlet.

Gall bladder: sac located in the visceral region behind the liver's right lobe that stores bile for secretion.

Gamete: mature reproductive cell, secondary oocyte ovum or sperm cell.

Ganglion: cluster of nerve cell bodies located outside the CNS.

Genitalia: internal or external reproductive organs.

Gestation: period of development between conception and birth.

Gland: organ or structure in the body that secretes or excretes substances.

Glomerulus: cluster of capillaries in the kidney.

Gonad: sex gland where reproductive cells are formed.

Haematopoiesis: production of blood cells, it occurs in the bone marrow.

Haemoglobin: quaternary protein found in red blood cells, it contains iron and is involved in the transport of oxygen by red blood cells.

Heart: muscular organ that pumps blood around the body.

Hormone: chemical substance secreted by an endocrine gland to trigger or regulate a group of cells or organ.

Humerus: long bone of the upper arm. It connects the shoulder and the elbow.

Humoral: relating to body fluids or substances, especially serum.

Hypothalamus: region of the brain located in the diencephalon that synthesizes ADH and oxytocin.

Ileum: distal part of small intestine from jejunum to caecum.

Inguinal: referring to the groin.

Insulin: hormone produced by the pancreas that enables the body cells to use glucose. It is secreted in response to elevated blood glucose levels.

Intestine: region of the GI tract that extends from stomach to anus.

Intima: refers to an innermost structure.

Jejunum: one of three portions of the small intestine, it connects the duodenum with the ileum.

Joint: a fibrous, cartilaginous or synovial connection between bones.

Kidneys: pair of urinary organs on the dorsal part of abdomen responsible for filtering waste out of the blood.

Labia: external part of female genitalia.

Lacrimal: referring to tears.

Larynx: (pl. larynges) voicebox; links pharynx to trachea.

Leucocyte: white blood cell, defends the body against pathogens.

Ligament: white fibrous tissue that connects bones.

Liver: large organ located in upper right abdomen, divided into four lobes.

Lobe: a portion of any organ.

Lobule: small lobe.

Lumbar: area of the back between thorax and pelvis.

Lungs: pair of organs of respiration, located in lateral cavities of the chest.

Lymph: fluid of the lymphatic system.

Lymph node: small, oval structures that filter lymph, aids haematopoiesis.

Lymphocyte: type of white blood cell, namely B-cell, T-cell and natural killer (NK) cells.

Malleus: hammer-shaped bone, one of three found in the middle ear.

Mammary: referring to the breast.

Manubrium: upper part of the sternum.

Mediastinum: middle region of the thorax between the pleural sacs that contains the heart (but not the lungs).

Medulla: inner portion of an organ; opposite of cortex.

Membrane: thin layer.

Mesencephalon: see *midbrain*.

Metacarpals: bones of the hand between the wrist and fingers.

Metatarsals: bones of the foot between tarsal bones and toes.

Micturition: see *urination*.

Midbrain: (mesencephalon) region of the brain stem that connects cerebrum with the pons and cerebellum. Involved in motor coordination.

Muscle: fibrous structures that contract producing movement.

Myocardium: thick contractile muscle surrounding the heart, forming the heart walls.

Nares: nostrils.

Necrosis: an undesirable form of cell death usually caused by factors external to the cell, such as trauma, toxins or infection. Gangrene is an example of a dangerous accumulation of necrotic tissue.

Nephron: structural and functional unit of the kidney.

Nerve: fibre that rapidly conveys electrical impulses between the CNS and PNS.

Neurone: nerve cell.

Neutrophil: most abundant type of white blood cell that destroys bacteria and cellular debris.

Oesophagus: muscular tube that transports food from the mouth via the pharynx into the stomach.

Olfactory: referring to sense of smell.

Ophthalmic: referring to the eye.

Osmosis: movement of water from an area of high concentration to an area of lower concentration across a selectively-permeable membrane. Movement continues until equilibrium is reached.

Ossicle: small bone (especially of the ear).

Osteoblast: bone-forming cell.

Osteoclast: cell that removes unwanted bone.

Ovary: (pl. ovaries) female reproductive gland located on each side of the lower abdomen.

Palate: roof of mouth.

Pancreas: secretory gland in the epigastric and hypogastric regions.

Parotid: located near the ear.

Patella: the knee cap, it is the largest sesamoid bone in the body.

Pectoral: referring to the chest or breast.

Pelvis: lower part of the trunk.

Pericardium: fluid-filled sac surrounding the heart and proximal ends of the aorta, venae cavae and the pulmonary artery.

Peripheral nervous system (PNS): the neurones connecting the CNS with the rest of the body.

Peritoneal cavity: space between the two layers of the peritoneum. It contains peritoneal fluid.

Peritoneum: double-layered membrane: the visceral peritoneum covers abdominal organs, the parietal peritoneum lines the abdominal wall.

Phalanx: a tapering bone forming the fingers and toes.

Pharynx: tubular passageway from the base of skull to the oesophagus.

Phrenic: referring to the diaphragm.

Pia mater: innermost membrane of the brain and spinal cord.

Pituitary gland: endocrine gland attached to the hypothalamus, it stores and secretes hormones.

Plantar: refers to sole of foot.

Plasma: fluid content of blood and lymph.

Platelet: small blood cell that forms part of blood, necessary for blood coagulation.

Pleura: thin serous membrane of the lungs.

Plexus: joining of a network of nerves, lymphatic vessels or veins, for example, choroid plexus is joining of network of blood vessels in the brain.

Pons: region of the brain between the medulla and cerebellum.

Popliteal: back of the knee.

Pronate: turning the palm downward; opposite of supinate.

Prostate: male gland surrounding the neck of the bladder and urethra.

Pupil: circular opening in the iris of the eye that allows light to enter the eye.

Radius: longer bone of the forearm, located on the lateral side of the ulna, it connects the elbow with the thumb side of the wrist.

Reflex: an involuntary (unconscious) action.

Renal: referring to the kidneys.

Rotation: a circular movement around a fixed point, such as a joint.

Scrotum: external pouch containing the testes of male genitalia.

Semen: male reproductive fluid consisting of sperm cells and supporting secretions.

Sesamoid bone: a bone embedded in a tendon (for example, the knee cap).

Skull: see *cranium.*
Sphenoid: wedge-shaped bone at base of skull.
Spleen: highly vascular organ of the lymphatic system located in the upper left quadrant of abdomen between the stomach and the diaphragm.
Stapes: stirrup-shaped bone, one of three in middle part of ear.
Sternum: long, flat bone forming middle portion of thorax.
Stomach: major digestive organ, located in the upper right quadrant of the abdomen.
Supinate: turning palm of hand upwards; opposite of pronate.
Synapse: connection point between adjacent neurones.
Synovial: the most common and most movable type of joint.
Systole: contraction of heart muscle, chambers empty of blood; opposite of diastole.
Talus: ankle bone.
Tarsus: instep.
Tendon: fibrous connective tissue that attaches muscle to bone.
Testis: (pl. testes) male gonad that produces sperm cells.
Thyroid: gland that secretes thyroxine, located at the front of the neck.
Tibia: shin bone, the larger of the two bones in the lower leg, it connects the knee with the ankle.
Tongue: main organ of taste since its upper surface possesses many taste buds. It is a muscle and is located in the floor of the mouth.
Trachea: tube in the respiratory tract, extending from larynx to bronchi.
Ulna: bone on the inner side of the forearm.
Ureters: pair of tubes that transport urine formed in the kidneys to the bladder (one from each kidney).
Urethra: small tube that drains urine out of bladder during urination.
Urination: (micturition) voluntary process that empties the bladder of urine.
Urine: waste liquid produced by the kidneys, collected and stored in the bladder and excreted from the body through the urethra.
Uterus: hollow organ of female reproductive system.
Vagina: canal in the female reproductive tract extending from cervix to vulva.
Valve: structure that permits fluid movement in only one direction, preventing back-flow (especially in heart and blood vessels).
Vein: vessel carrying blood towards the heart.
Vena cava: (pl. venae cavae – superior and inferior vena cava) vessels entering the right atrium of the heart carrying deoxygenated blood.
Venous return: rate of blood flow back to the heart in the veins, it limits the rate of cardiac output.
Ventricle: small cavity; there are several ventricles in the brain; also describes the two lower chambers of the heart.

Venule: small vessel connecting a vein with a capillary.

Vertebrae: (sing. vertebra) the 33 bones that make up the vertebral (spinal) column.

Viscera: internal organs.

Zygote: a single diploid cell resulting from the fusion of sperm and ovum.

PARAMEDICS! TEST YOURSELF IN PATHOPHYSIOLOGY

Katherine Rogers, William Scott, Stuart Warner and Bob Willis

9780335244515 (Paperback)
October 2011

eBook also available

Looking for a quick and effective way to revise and test your knowledge?
This handy book is the essential self-test resource to help paramedics revise and prepare for their pathophysiology exams. The book covers most common presentations seen in the paramedic practice and includes over 250 questions and 60 glossary terms in total.

Key features:

- Organised into body systems chapters
- Includes a range of question types
- Provides a list of clearly explained answers to questions

www.openup.co.uk

 OPEN UNIVERSITY PRESS
McGraw - Hill Education